MAUSER
Rifles & Pistols

Originally published in 1958

ISBN: 978-1-940001-24-1

Published by Sportsman's Vintage Press
2014

A NATIONAL RIFLE ASSOCIATION
LIBRARY BOOK*

Mauser Rifles and Pistols

*A National Rifle Association Library Book means one recommended as the standard work on the subject by the National Rifle Association of America.

WILHELM MAUSER
1834-1882

PAUL MAUSER
1838-1914

AN NRA LIBRARY BOOK

MAUSER RIFLES *and* PISTOLS

W. H. B. Smith

Author of

Small Arms of the World

Mannlicher Rifles and Pistols

Walther Pistols

NRA Book of Small Arms:
Vol. I—Pistols and Revolvers
Vol. II—Rifles

Gas, Air, and Spring Guns of the World

CONTENTS

Mauser Rifles

Mauser Pistols

COUNTRIES USING MAUSER RIFLES

Argentina, pattern 1891, caliber 7.65 mm, .301 inch.
Belgium, pattern 1889, caliber 7.65 mm, .301 inch.
Bolivia, pattern 1891, caliber 7.65 mm, .301 inch.
Brazil, pattern 1904, caliber 7.00, .276 inch.
Chile, pattern 1904, caliber 7.00 mm, .276 inch.
China, pattern 1893, caliber 7.00, .276 inch.
Columbia, pattern 1891, caliber 7.65 mm, .301 inch.
Czechoslovakia, pattern 1924, caliber 7.9 mm, .311 inch.
Ecuador, pattern 1891, caliber 7.65, .301 inch.
Germany, pattern 1898 modified, caliber 7.9 mm, .311 inch.
Luxembourgh, pattern 1896, caliber 6.5 mm, .256 inch.
Mexico, pattern 1902, caliber 7. mm, .276 inch.
Peru, pattern 1891, caliber 7.65 mm, .301 inch.
Portugal, pattern 1904, caliber 6.5 mm, .256 inch.

Note: Portugal adopted the modified Kar. 1898 K during World War II, caliber 7.9 mm, .311 inch.

Persia, 1889, 1891 and 1902 models in rifles and carbines, various calibers, mostly 7.65 mm, .301 inch.

Poland, pattern 1904 modified, caliber 7.9 mm, .311 inch.
Spain, pattern 1896, caliber 7.00 mm, .276 inch.
Sweden, pattern 1906, caliber 7.65 mm, .301 inch.
Turkey, pattern 1905, caliber 7.65 mm, .301 inch.
Uruguay, pattern 1895, caliber 7 mm, .276 inch.
Yugoslavia, pattern 1899, caliber 7 mm, .276 inch.

Besides the above list of genuine Mauser rifles made in Germany, Mausers are also made in huge numbers and in any military caliber by the Steyr Works in Austria, the F. N. Works in Belgium, the Brno Works in Czechoslovakia for export. The Radom and Warsaw Arsenals in Poland, the Oviedo Arsenal in Spain and numerous Chinese plants all make copies or slight modifications of the German Kar. 98.

Note: All later patterns of Mauser rifles both in calibers listed and also in the standard German Service 7.9 mm caliber will be found in the countries above. The most common are the slightly modified Kar. 98 types made to German specifications in Belgium, Czechoslovakia and Spain.

COUNTRIES USING MAUSER PATTERN RIFLES, WITH MODIFICATIONS MADE IN COUNTRY OF MANUFACTURE:

Italy 1891, Carcano, caliber 6.5 mm, (2.56-inch) and later 7.35 mm model 1935. Mauser-type bolts.

Japan, Year 38 Pattern 6.5 mm, (2.56-inch) and later 7.7 mm, Pattern 99.

United States, 1903-06, Springfield, caliber 7.62 mm, (.30 inch). Mauser-type bolts, receivers and magazines.

AUTHOR'S NOTE

I

This book has been prepared directly from the records of the Waffenfabrik Mauser, from reports and publications by directors of the Mauser Werke A.-G., (Firearms Factory), from the Mauser memorial and records of R. H. Korn, who was Paul Mauser's personal patent attorney, from the records of Ludwig Loewe and Co. of Berlin who were Mauser partners for many years, and from accounts of the Württembergische Vereinsbank of Stüttgart, financial partners and bankers for Mauser.

All photographs and drawings and all specifications except as noted are from the original drawings, photographs and specifications prepared by the Mauser organization for their official reports or for their international sales organizations.

Wherever possible direct translations from the pertinent German records have been retained in their essence, being expanded, edited and elaborated only where necessary to amplify the original text.

II

This book, while essentially an historical hand book for gun enthusiasts and for the millions of soldiers who possess European Mauser rifles as trophies, also has a secondary purpose. That secondary purpose is to present a detailed study of the advance of German small arms development through the years, so politicians and economists and statesman will have available an integrated study of *one facet* of militaristic evolution which has never before been presented.

This book is confined entirely to the facts of the Mauser history and of the arms Mauser developed; but its implications in world affairs are there for any analysts who are in a position to use them.

As a history of the rise and development of the Mauser line of weapons, this work is intended to satisfy all except the ultra-technical.

III

Very few people in civil life have any realization of the tremendous part the Mauser organization played in world affairs from the date of its inception in 1871 to the close of World War II.

As a result of a planned German military policy, a policy which began with the very founding in 1864 of Germany as the Nation we knew in the Twentieth Century, the Mauser organization was used internationally to disseminate German ideas and German ideologies.

Germany infiltrated the entire South American continent and much of China by recognizing the elementary fact that whoever armed the police and military organizations of those countries, automatically exercised considerable control over their politics and policies. In world areas where police and military groups dominate, those who are in a position to provide arms and equipment and new military techniques have always been able to achieve a measure of power entirely out of relation to either the number of agents employed or the extent of the business they have done with those countries.

IV

As this book is written, the Mauser factories in Germany (in U. S. and British areas!) for the second time in less than half a century have been compelled to stop manufacture of weapons. But—again for the second time in that period—their *manufacturing potential* has not been seriously impaired. Meanwhile, in Russian areas they are reported in full operation!

Very early in the history of the Mauser organization, contacts and affiliations were made with the great Austrian Steyr Armory for the manufacture of German designed weapons in Austria. At a slightly later date similar arrangements were made with arms companies in Belgium, notably at Lüttich.

At the close of World War I, when the Mauser factories were in Allied hands, the great Fabrique Nationale at Herstal, near Liege, Belgium, undertook to manufacture German Mauser rifles to meet the legitimate police and military requirements of Central and South America and of much of the Orient.

It is noteworthy indeed that all these Mauser rifles took the general pattern of the official German Gewehr 98! The differences in manufacture and in manufacturing methods were so slight that a very high degree of interchangeability of parts has always existed among Mauser rifles, *wherever made, and in whatever caliber made,* throughout the world. Not only were the receivers (the central forging which is the heart of any rifle and which houses the operating mechanism, and into which the barrel is screwed) of the same design and general length and weight, but all the military *cartridges* designed for the weapon were similar enough in overall dimensions that comparatively few changes were necessary in machinery and manufacturing to convert the Central and South American calibers to those of the German standards. *This condition still exists.*

When arms factories were set up for the manufacture of rifles at Brno in Czechoslovakia, again not only were German measurements and requirements instituted, but the actual German military cartridge caliber was retained! These Mauser rifles of Czech manufacture are among the finest known. They vary from the German only in *very slight* details. These factories also entered world commerce to provide military and police arms wherever required; and their products will be found throughout Central and South America and the Orient.

When arms factories for Mauser rifle manufacture were established at Warsaw and at Radom in Poland, again the arm manufactured was in all essentials the standard German army rifle using the standard German rifle cartridge!

It must be remembered that with the sole exceptions of the United States, which uses the Garand M1 semi-automatic rifle and the American Springfield (which is an improved Mauser rifle); Great Britain which uses generally the Lee-Enfield rear lug system (together with huge numbers of Mauser type American made Model 1917 rifles); France which adopted a modified Mauser pattern in 1935; and Russia which uses its own Tokarev semi-automatic rifle and its Nagant bolt action rifle, practically every country in the world today is armed officially with weapons of German Mauser or Austrian Mannlicher design. Italy, which employed at the beginning of World War II rifles with the Mannlicher magazine, also used a modified Mauser bolt.

In this day of atom bombs and rockets and flame throwers, it is easy to overlook the political implications of a world system which permits a militaristic nation to provide arms to the police and military authorities of smaller nations. Where those arms and their replacement parts and ammunition go, there too go instructors, commercial agents, and exponents of the political and military ideologies of the country providing those weapons.

Regardless of the implications of the *mass-destruction weapons* now available in the world, they are likely to be used only under the direst of circumstances. Military, police and sporting arms however, will continue to be, as they have in the past, of supreme importance to the individuals directly concerned. Thus any organization selling and distributing weapons and techniques to the police or the military in any small nation is much more likely to dominate that nation's policies and thought than is one which depends entirely on *standard* commercial, economic or cultural contacts.

As this is written, most of the great German, Austrian, Polish and Czech arms plant capable of manufacturing Mauser rifles or replacement parts for them are under direct or indirect Russian control or influence. Russian trade missions are abroad in every Central and South American country. At this juncture only Russia and Belgium are in a position to supply or maintain the police and military equipment of our Central and South American neighbors. And Belgium can't begin to undertake to fill the legitimate demand.

A move is currently on foot to assist our neighbors by re-arming them with standarized United States equipment. This move is meeting political opposition. If we fail to take full advantage of this opportunity at this time, those nations have no recourse but to turn to Russia for adequate essential small arms requirements.

The effective spread of German militaristic ideas anywhere the Mauser organization went is the best evidence of what *can* happen in the event Russia is able to follow in the footsteps of Germany on an international scale.

HISTORICAL FOREWORD

The following history is the result of original research and has not heretofore appeared in English except as noted.

I

This is the fantastic story of an application of the elementary turning door-bolt which made *military* history. It is the story of how the German Nicholas Dreyse recognized and utilized the idea of applying the push-and-turn-down principle of the common door-securing bolt to lock a cartridge in the firing chamber of a rifle. So secure was this new lock, that an entirely new form of military rifle was evolved which literally changed the face of war.

It is the story of how two lowly German gunsmiths, Paul and Wilhelm Mauser, seized upon and perfected the elementary application of Dreyse and used it in conjunction with a newly developed metallic cartridge to produce the first truly successful German rifle.

From that point on the simple story of the door-bolt rifle enters the ramification of world history. From his first crude design, Paul Mauser evolved the bolt action magazine rifle bearing his name. This rifle was to remain the standard arm of the German forces through two World Wars and was to become the standard military arm of practically all the smaller nations of the World, while its modifications were to be adopted by most of the remaining nations.

Paul Mauser devoted his entire adult life to the further development and evolution of military arms in the interests of his Fatherland. The effect those weapons have had on history is much too great to be assessed in a work of this nature. But there is no corner of the globe in which power on the one hand and violent death on the other has not followed in the footsteps of the Mauser salesman. The sale of sporting Mausers has always been purely incidental, they were merely modified military arms.

II

Peter Paul Mauser was born on the 27th of June in the year 1838 in the little town of Oberndorf on the banks of the beautiful German river Neckar. Peter Paul was the youngest of the thirteen children Merira Agatha Heím Mauser bore to Andreas Mauser of Sondheim.

Andreas Mauser himself was a master gunsmith in the Government Firearms Factory housed in the former Augustine Cloister at Oberndorf. This factory was originally established by King Frederick I by a Cabinet Decree of July 31, 1811. Originally located partly at Ludwigsburg and partly in Christophsthal, it was transferred to Oberndorf because of the fine facilities available there. The old Augustine Cloister, built from 1775 to 1788, had been vacated by virtue of the secularization in the year 1806; and in accordance with the history of such institutions from the earliest times, was found to be so well constructed and so ideally situated that its conversion to an arms factory proved to be a very logical move. It is indeed a strange coincidence that even in England, as far back as the year 1539, a convent which had stood since 1293 was taken over by Henry VIII; and under the name of "The Minories" became outstanding in the gun manufacturing trade in England. Throughout the course of recorded history men and women of peace have selected beautiful sites and constructed fine buildings only to have them eventually seized and turned to factories for war.

Peter Paul, to give him his full name, his next older brother Wilhelm, as well as five of the older sons, were trained by their father in the line of his interest. When he was only 12 Paul was already busy at the gunsmith's work bench. Throughout all the days of his youth, his family was financially poor and his meager earnings helped support the large Mauser family. In 1852 when Peter Paul was graduated from high school, his future life work was already decided. He joined his father and brothers in the Government Firearms Factory and soon attracted attention by his unusual ability to develop new methods of work, new short cuts in manufacturing processes, and specialized tools which enabled him to produce faster and better than his older bench mates.

Europe was then, even as now, a Continent seething with discontent, fear and suspicion, and Germany employed a rigid system of compulsory military service. In 1859 Peter Paul was called up for military duty.

Since military systems throughout the world have always been noted for their prodigal waste of talent and ability, it is not surprising that Germany assigned Peter Paul, brillant small arms mechanic, to duty as an artilleryman at the arsenal at Ludwigsburg. In later years Peter Paul himself credited his study of the models he saw at Ludwigsburg, particularly of the new but highly imperfect breech loaders, with starting him on the development of rifles which brought fame and fortune to him and military might of ephemeral quality to his country. He lived to see his Germany rise in military might; he was fortunate enough to die before its star fell in disastrous defeat.

By December of 1859, Peter Paul had so impressed his immediate superiors that he was placed on inactive status and assigned to the Royal Firearms Factory at Oberndorf. Time had made inroads into the Mauser family. Father Andreas had died. Josef and Heinrich, the two oldest brothers, had married and had their own families to think about. Another brother Franz, had emigrated to America where he was to become an employee of the Remington organization, one of the great pioneers of American firearms.

And so it was that Peter Paul, faced with the desire to turn his creative energy to practical purposes that he might earn a better livelihood for himself and for his family, approached his brother Wilhelm, who was two years his senior, and asked him to work with him on the creation of a new gun in the evening hours after their day's work in the factory was done. Then, as in later years, the factory was a hive of industry. The hours were long, the labor was gruelling. Only a person with dogged courage, stamina and determination could summon up enough energy at the end of a hard days work there to put in long hours of home work, whatever the potential return. The energy, will and drive of Peter Paul Mauser were elements in his success throughout the entire course of his life. Whatever one may feel about the *directions* his boundless energy took, there can be no argument about its drive and intensity.

As the older of the two brothers, Wilhelm automatically assumed his father's place as negotiator in the contacts and dealings the Mauser brothers had with the factory. From the first his health was delicate and this, together with a natural ability as a salesman, spurred him on to take active charge of the presentation of their interests, while Peter Paul did most of the actual experimental, technical and development work.

The experiences of Peter Paul while in the artillery influenced his thought to such an extent that his first invention was a small breech loading cannon and a special steel projectile of unusual design for it. While he credited Wilhelm with part of the collaboration on the cannon itself, Peter Paul alone was the developer of the ammunition. Throughout his life he claimed to be the sole creator of the ammunition used and developed for Mauser weapons all over the world. This fact is of great significance in the development of firearms. The ability to develop not only the mechanical principles of a weapon but also the ballistically correct ammunition for the mechanical principles involved is a tremendous factor in the successful production of firearms.

While this breech loading cannon was of general interest and was later preserved at the Royal Army Museum at Stüttgart, the difficulties of producing it with any hope of profit very soon taught the brothers a much needed lesson. Since their finances were limited, their activities must needs be funneled into channels where large sums of money were not involved in experimentation, initial production and selling.

Both brothers of course were thoroughly familiar with the Dreyse Needle Gun then in German military use, and after its outstanding military success at Alsen in 1864, they turned their combined efforts to ways to improve the locking and functioning of this new military arm, which was based on the locking principle of the elementary turning door-bolt.

For a time there was a dangerous rift in the friendly and fraternal relations of Wilhelm and Peter Paul, and for several years they went their separate ways. Finally Peter Paul succeeded in creating a turning-bolt lock which by a simple cam action during opening and closing of the breech mechanism would cock the mainspring.

Wilhelm was so impressed with this development that he again entered into business relations with his younger brother, and he was so successful and so forceful in presenting this new development to the military authorities that the government granted the two brothers several hundred florins with which to purchase machinery for further development work. The brothers later were able to repay this loan, but without it they could never have completed the experimental work so essential to the development of the arm on which they were engaged.

The first Mauser developments were connected with rifles using the needle principle, in which the needle at the forward end of the striker was driven forward through the powder charge to hit and discharge the percussion or priming charge at the base of the bullet. Soon, however, they produced an advanced form of arm using a needle-percussion action.

The army of their native Württemberg had but recently been equipped with Minié Rifles; and as the investment had already been made in those arms, that government was no longer interested in a new rifle, even though it was an admittedly superior design. The financial commitment was too great to warrant a changeover.

The brothers next turned to the Royal Prussian Ambassador at Stüttgart with their new arm. That official, impressed by and glorying in the Prussian successes in battle with the Dreyse Needle Gun, decided arbitrarily that the Dreyse was so excellent that no change could even be considered.

Undaunted, the brothers next approached the Austrian Ambassador, who was considerably more receptive. He forwarded their new rifle to Vienna for tests, and that action started a new chain of events in the lives of the Mauser brothers.

III

Few arms enthusiasts know that the first Mauser rifle patent was taken out *in the United States.* That rather strange development was a direct outgrowth of their initial presentation of the new weapon to the Austrian Ambassador.

At that period many countries in Europe and the Orient were considering arming with the then famous American Remington rolling-block rifles. That was the heydey of breech loading rifle development in the United States. Peabody had developed a fallingblock breech loader which was being used, modified or adapted in great numbers throughout Europe; the Winchester repeater was making itself felt as a force in Turkish military life; and from Norway to Egypt, from Spain to Turkey, and even in far away South America and China the Remington rolling-block was being considered for military service.

Representatives of American firearm manufacturers were persona grata in *all* corners of the world. It is not strange then that Mr. Charles Norris of the Remington Company should be calling on the Austrian Minister of War at Vienna. Austria had but recently changed over to the Wänzl rifle and the manufacture was in such an advanced stage that, like Württemburg, Austria could not afford to bypass the Wänzl even in the face of an admittedly superior design. It was at the War Ministry that Norris first saw the Mauser rifle.

The Austrian War Minister was quite frank with Norris in pointing out to him that only the financial commitments already involved in the Wänzl changeover prevented Austria from adopting this new German design. Norris, a true Connecticut Yankee of story-book type, at once grasped the possibilities of a fine business opportunity at practically no cost or risk. The French assured him they would be interested in a system to convert the Chassepot to a metallic cartridge rifle; and Norris travelled to Oberndorf with the sole idea of tieing the Mausers up in a contract which would give him control of a Chassepot conversion. His classic contract and his own story will be found at the close of this chapter.

Norris hired the Mausers to go to Lüttich in Belgium, then the seat of firearms design in Europe, where all facilities necessary for further development would be readily accessible, to perfect the design for him. Norris also stipulated that patents should be taken out in his name and that the Mauser brothers were to receive a royalty on the proceeds of weapons sold. Designers and inventors from the earliest of times have been subject to this type of promotional contracts, which very seldom work in actual practice. The rifle, called the "Mauser-Norris" was duly patented in the names of Norris and of the two brothers in the United States. The Remington firm, justifiably incensed at the action of their European representative, were sold an interest in the Mauser contract by Norris. They failed to push the new bolt action rifle, however.

At that time, the Royal Firearms Factory at Oberndorf, being in the process of reconverting from the Minié muzzle-loading rifle to the more successful breech-loading needle gun, was forced to lay off a great number of workmen, including the Mauser brothers. And so it was that early in 1867 the Mausers, glad of any work which would pay their expenses, moved to Lüttich in Belgium. For two years Peter Paul and Wilhelm worked incessantly at further developing their rifle. Their stay was mentally stimulating at least, as it brought them in contact with many of the outstanding experts and manufacturing geniuses in the European armament industry then situated in Lüttich.

IV

By the Fall of 1869, Norris failing to meet his obligations and provide financial support to the Mauser brothers, the two were compelled to return to their family home at Oberndorf where they set up to manufacture rifles in a small workshop in the home of Peter Paul's father-in-law.

Before leaving Lüttich they had insisted that Norris submit their rifle to the Royal Prussian School of Riflemanship. The results of the test there were so impressive that Wilhelm was invited to the Arsenal at Spandau. The Institute at Spandaü later produced the Royal Rifle Testing Commission, a body to which Paul Mauser submitted all his rifle designs from that period on. Throughout the course of his life he was always proud, in the way that only a German militarist could be, of his connection through the years with the top rifle authorities of Prussia. What part Peter Paul himself played in later years in the planned system of distribution of his arms as instruments for spreading German thought and military policy throughout the world, it is difficult indeed to say. But the record does show that all his efforts and all his abilities were harnessed to the one idea of developing standardized equipment and munitions.

The scond of December 1871 was a red letter day in the lives of the brothers Mauser. It was on that day that their rifle was adopted as the first official

German service metallic cartridge rifle, even though the Testing Commission required a number of changes in the basic design, particularly in the safety device.

Peter Paul, enthralled at the prospect of the honor of having his rifle adopted, worked day and night to perfect a new safety lock; and on the 14th day of February 1872 submitted two systems to Spandau for their selection. Like practically every other mechanical development of Peter Paul Mauser, this safety was so fundamentally correct that it was never possible to do more than refine it. Even today it is a characteristic feature of the finest Mauser rifles.

While the genesis of the turning-bolt action lock is usually credited to Dreyse, and the overall form of that first Mauser rifle is often thought to resemble closely that of the French Chassepot, the truly revolutionary features in the design are strictly those of Peter Paul Mauser.

Out of that elementary house door-bolt he produced a unit which was self-cocking, had a distinctive bolt head, utilized an elastic extractor, incorporated an effective ejector, and embodied the famous cam operation for giving "primary extraction" to loosen swollen cartridge cases, without which a truly successful military rifle could not function.

Thus this first rifle, officially listed as the "Model 71", was really not produced until 1872.

The Prussian Army promptly supplied a substantial order and the brothers Mauser set up a temporary workshop in a little building in Oberndorf. That first plant employed 50 people on its opening, utilized several special machines developed by Paul Mauser, and was powered by two movable steam engines. Soon the plant grew to utilize the energies of 100 workers.

By now the brothers knew that success was within their grasp and they unhesitatingly committed themselves to the erection of a new plant of their own on the height of Oberndorf, the plant later known as the "Oberes Work". In the summer of 1872 they laid the foundation stone of this new plant, and day by day in proud anticipation they watched it grow.

Again an ill wind blew, the kind of wind that blows down and destroys weaker men than the brothers Mauser. Hardly had the building been completed on the heights then a great fire occurred; and on the 20th of August 1873, the building was badly gutted by fire.

With truly indomitable will and courage, they passed off the severe blow and turned their energies to refitting the factory. Just eight weeks from the day of the fire the plant was able to resume operations.

By now the High Command of the Württemberg army was convinced not only of the quality of the rifle but also of the ability and stamina of the Mauser brothers. They therefore offered a contract for approximately 100 thousand rifles and carbines of the Model 71 to outfit their army. By this time the pattern of German military thought with its long term view of ultimate world power was taking form. The new German Confederation was shaping and the Germany of Kaiser Wilhelm and later of Adolph Hitler was beginning the climb to the military heights which was to produce two generations of World chaos. The Württemberg Government offered the Mauser brothers the Government Firearms Factory for a price of 200 thousand florins. Financing this project, a gigantic one from the standpoint of the Mauser brothers, was simplified by the government-arranged participation of a banking concern which could offer the necessary financial guarantees to the government.

And thus, from a small struggling business, the destinies of the two Mauser brothers transformed their organization overnight into a great industrial concern. The newly organized Württembergische Vereinsbank of Stüttgart set up the Mauser Brothers and Company as a corporation and entered as a partner with an investment of 800 thousand marks. On the 20th of February 1874 the Chamber of Deputies officially signed the contract which became legal two weeks later transferring the Government Firearms Factory to the apparently non-military Mauser Brothers and Company Incorporated. An interesting sidelight on this development is that Paul Mauser himself had set up and was operating his arms factory in the ancient Augustine Cloister by the middle of December, 1873, and had established his home nearby—a good three months before the matter officially came up before the Chamber of Deputies. Wilhelm, meanwhile, had moved to the height where he supervised the Oberes Work.

In spite of the fact that the machinery at the former Government Firearms Factory was obsolete and had to be replaced with completely new equipment, the driving efforts of the two brothers resulted in completion of the Württemberg Government order in 1876, nearly six months ahead of schedule. Minor improvements were made in the Model 71 and Wilhelm obtained an order from the Chinese Government for 26 thousand rifles which was promptly filled.

For the next few years only the cooperation of the German military acting through their financial backers, kept the Mauser brothers in business. Wilhelm in particular, haunted the chancelleries of Europe looking for new business. Finally, in 1881 after intense competition both from the standpoint of quality of the arms and of the politics and economics always involved in military deals on an international basis, the Mausers succeeded in having the Serbian government adopt their arm. They were given an order for 100 thousand infantry rifles of caliber 10.15mm with an improved version of the famous Mauser lock. This order was the swan song of brother Wilhelm. Long ill and in intense suffering, he lived only long enough to see the Serbian order underway. On his death on the 13th day of January in 1882 the burden of support of Wilhelm's family was taken over by Peter Paul. On his head and broad shoulders too fell complete charge of the operation of the plant which had contracted to complete the Serbian government order by the Spring of 1884. With all his duties and all his responsibilities, Peter Paul Mauser still found time to work steadily on new inventions, working with an indescribable energy which left him no time for social life.

He developed a breech loading pistol and a revolver which were patented in most of the great nations of the day. The introduction of the inferior Models 73 and 79 Service revolvers met with very little success in Germany, and one-hand weapons as such fell into military disrepute until the later development by

Paul Mauser of the semi-automatic pistol. The failure of the Government revolvers as efficient weapons served to impede the military sale of the Mauser pistol and revolvers, although his arms were of superior design.

V

The success of the Spencer and Henry repeating rifles introduced in the American Civil War, and of the Winchester rifle which grew out of the Henry and was used with terrible effect by the Turks against the Russians, forced German military officials to recognize the need for a magazine rifle which would give increased infantry firepower. Peter Paul Mauser, being very close indeed to the Rifle Testing Commission, was of course thoroughly familiar with the desire of the military for such a development. He created several rather interesting designs in box loading magazines which were the beginning of his later successful types which became world standard. The successful ones are all listed in the body of this work. There were few types, however, with which he did not experiment.

The Henry-Winchester system of carrying cartridges in a tube below the barrel where they compressed a spring which thrust them successively back into a carrier for individual loading in the chamber, was an immediate success. Hence it was promptly picked up and utilized in experimental rifles in most of the important military nations in Europe.

In the Fall of 1880, Mauser applied this cartridge carrying principle to his original Model 71 Single Shot Rifle, a very important development as it permitted the use of standardized machinery and enabled the conversion of the single shot design to repeating rifle design at a minimum cost.

One of the proudest moments in the life of Peter Paul Mauser was a day in September of 1881 when at the Württemberg, Industrial Exposition he was permitted to demonstrate his new magazine rifle to His Imperial Majesty Wilhelm I (1799-1888). It made such an impression on His Majesty, that a test by the Rifle Testing Commission was expedited. That trial was so successful that Mauser received an order for 2000 test weapons.

Those 2000 rifles were put into the field by the Prussian High Command for complete testing under field conditions. The rifle was shortly thereafter adopted under the official designation of "Infantry Repeating Rifle Model 71-84, caliber 11mm."

Serbia at that time was a hot bed of militarism. Paul Mauser's contacts were so good that he was able to obtain from the Serbian War Office an order for 4000 rifles and 4000 carbines, identical with the German type, but adapted to a special 10.15mm caliber cartridge used by the Serbians.

The Mauser Company about this time found it essential that they obtain additional work to keep the plant busy and they obtained military cooperation in the form of an order for 19,000 M71-84 rifles to arm the Württemberg Armies. The Bavarian and Prussian Governments manufactured the M71-84 in their respective Government Factories at their own expense, paying the Mauser corporation a royalty on each weapon.

The next forward step in the fortunes of Mauser Brothers & Company, was when it became a stock company on the First of April, 1884. The Württembergische-Vereinsbank placed its assistant director Alfred Kaulla on the Board of the Mauser Company to handle financial details, and left Peter Paul Mauser in sole charge of technical developments. In later years Kaulla was the true financial genius of the organization.

VI

This was a period of change in Europe, a period of uncertainty in military circles, a period when technical developments in the field of explosives in particular, was so encompassed in rumor and experiment that each nation hesitated to change its weapons for fear of a giant technical stride which would at once invalidate their investments.

Mauser himself conducted long series of tests to determine the most efficient caliber using the black powder with which he was familiar. He finally established that a 9.5mm caliber gave the best ballistics with the powder then available; and during this time he also doubled the size of the locking lugs on his bolts as a protection against stepped-up breech pressures.

Meanwhile the French had put into use the development of the new smokeless powder satisfactorily produced by their chemist Vielle. This new powder permitted reduction of caliber to 8mm with improved ballistic performance to a degree which startled and worried all the other armies of Europe.

Mauser, receiving no encouragement from his own government at this period, went to England with his new rifle, and again received a very cold reception. His next stop was Constantinople and in 1886, when the Turkish army was still equipped with American Remington and with Martini-Henry Single Shot Rifles and a sprinkling of American Winchesters, his new development received the attention to which militarily it was entitled.

After a series of tests and the customary amount of juggling inherent in government arms deals at that period, the competitive contract was finally granted to Mauser. Just what part the great Berlin firm of Ludwig Loewe & Co. played in obtaining this order is difficult to assess; but when Mauser received his order for 500,000 rifles and 50,000 carbines, it developed that Loewe was a 50-50 partner in the order. They took over all the stocks of the Mauser Firearms Factory and also undertook a huge order (425,000 rifles) for the newly adopted German Model 1888 Rifle most of which were manufactured at the Loewe plant in Berlin. This left the facilities of the Mauser factory itself open for the production of Turkish weapons.

Paul Mauser was extremely bitter at the adoption by the German Rifle Testing Commission of the Model 1888 Rifle. This arm, which used a modification of the Mauser bolt, utilized the Mannlicher type of packet loading in which the loaded clip is inserted into the rifle as an integral part of the magazine. It was officially listed as the "Infantry Rifle Model 88" and popularly known as the "Mauser and Commission." Mauser forecast accurately the difficulties which would be encountered with this type of magazine loading, and time and military experience led to the eventual adoption of Mauser's own system of clip loading (in which the cartridges are stripped off the clip into the magazine), a loading system which is still the most common in world use.

The Turkish Government contract called for delivery of a total of 500 weapons per working day. As the rated capacity of the Oberndorf plant had never been higher than 250 rifles per day, this order called for more industrial genius, courage and organization on the part of the director. The lower works, the "Ausseres Werk," equipped with 100 horsepower steam engines in 1885, was enlarged and outfitted with new machinery in very short order.

VII

While all the manufacturing facilities of the plant were turned over entirely to this Turkish rifle, Mauser himself still found time to experiment and develop rifles to handle the new type of smokeless powder then being produced.

After considerable experimentation, he found that in spite of the French adoption of the 8mm, the finest ballistic performance with the new powder was procurable with caliber 7.65mm and he proceeded to design a rifle having a jacketed barrel and using a 5-shot magazine below the receiver which could be loaded through the top of the open action.

Thus when in 1889 the Belgian Government after intense and highly intelligent tests of all known types of small arms, determined to develop a new rifle to equip its army, Mauser offered his new rifle for test.

When it is remembered that for years the sole general industry of the Lüttich area had been weapons, and the added fact that the Belgian Government was noted for its probity, the acceptance of the Mauser rifle as the official Belgian arm establishes at once the high quality of the design and workmanship of Paul Mauser's product.

At this period there began the very close liasion between the arms manufacturers of Belgium and those of Germany which resulted in the development and use of strictly German types of arms for military use and for world wide sale since that time. The Belgian Ministry of War ordered several hundred thousands of these new Mauser rifles with the stipulation that they be manufactured by the Fabrique Nationale d'Armes de Guerre, at Herstal near Lüttich. This plant was established by a Lüttich syndicate with Ludwig Loewe & Co. of Berlin as a partner. These rifles were also made by the Fabrique d'Armes de L'Etat at Lüttich.

Thus was created the first truly successful *really modern* rifle.

With the Belgians completely rearming after these intensive tests and with the adoptions of smaller calibers by most of the European countries, the Turkish

Government felt impelled to change to the smaller rifles with their improved ballistic performance. Thus after Mauser had produced 220,000 of the rifles in caliber 9.5mm, Turkey formally adopted the new Model 90 in caliber 7.65mm Turkish.

By the Fall of 1893 Oberndorf had provided 200,000 rifles and carbines of this model to Turkey. Argentina even at that early date had an intelligent and aggressive military group which kept abreast of developments in arms. Argentina, therefore, approached Mauser to manufacture a rifle of the same general type as the Turkish but with a heavier bolt for their army. Since the comparatively small capacity of Oberndorf was engaged for several years ahead, Argentina placed its order for 180,000 rifles and 30,000 carbines to be manufactured at Berlin by Ludwig Loewe & Co. to Mauser's specifications. Again the part that Loewe played in obtaining this order is all too obscure, but a knowledge of the fundamental processes by which such deals are swung leads one to believe that while quality was all essential, the more prosaic matter of financial dealings also entered into the picture.

Mauser began direct negotiations with the Spanish Government in November, 1887, and endeavored to sell them his 9.5mm caliber rifle. The imminence of developments in the field of new explosives delayed the Spanish High Command from a commitment at that time; but in 1891 after his Belgian and Turkish models had demonstrated their worth, the Spanish gave Mauser an order for 1840 testing rifles of 7.65mm caliber as well as an order for 400 carbines of the same caliber for the Spanish Navy.

Paul Mauser visited Spain in 1892 after delivery of the trial orders and brought with him a rifle with improved magazine and designed to use a cartridge of 7mm caliber which he had developed for use with the new nitro powder. This new rifle had the now famous staggered magazine of Mauser construction with its fool-proof system of feeding which is the one in general military use in most countries today.

The Spaniards were so impressed with this new arm and its new cartridge that they not only placed an order with Mauser but also awarded him the Grand Cross of the Spanish Military Order of Merit, the highest decoration Mauser ever received.

Since Oberndorf under the Turkish contract could not manufacture anything but Turkish weapons during the life of their contract, the new Spanish weapons were again manufactured by Loewe, with the lone exception of 30,000 pieces which were manufactured in 1895 at Oberndorf on completion of the Turkish order. These Loewe manufactured Mausers of 7mm caliber were the famous rifles used by a small contingent of Spanish troops at San Juan Hill during the Spanish-American War, an engagement where despite the tremendous numerical superiority of the Americans, the Spaniards inflicted a terrifically high rate of casualties.

Two hundred and fifty-one thousand eight hundred Model 93 Rifles and 27,500 Model 93 carbines were delivered to Spain during this period, and the Arsenal at Oviedo in Spain was specially tooled up to manufacture Mausers.

The Spanish Mauser retains an enviable record in military circles to this day for its reliability and accuracy.

In 1893 also Mauser traveled again to Constantinople to submit his Spanish Model in person to the Sultan of Turkey. The Turks grasped at once the conspicuous improvements Mauser had made, and once again passed over the monetary and practical factors involved and halted production on a rifle in manufacture. By that time 280,000 of the Model 90 Rifles had been delivered and Turkey increased the order from 550,000 to 700,000 weapons including in it 201,100 rifles under the designation of Model 93, the design being the same, but retaining the caliber of 7.65mm Turkish and using a cut-off on the magazine which permitted single shot fire while holding the magazine in reserve.

VIII

On his return to Germany Mauser resumed with feverish activity his never-ending round of work. He developed a special rifle and carbine in 6.5mm caliber in which he interested the Swedish Government. In August 1894 they ordered 5,000 pieces and in June, 1895, another 7185 pieces. Again it is of significance that additional Mausers were manufactured not at the Mauser plant, but at the Carl Gustave Stads Firearms Factory at Eskilstuna from Mauser-designed machinery. Mauser received a royalty on each rifle made by the Swedish Government. However, in 1899, when the Swedes could not fill their own requirements, an order for 45,000 of these rifles was placed directly with Oberndorf.

IX

By the year 1894, the Germany Army had had sufficient experience with their Mannlicher type of rifle under field conditions to know that Mauser's original contentions had been correct—that under strenuous service the Mauser loading system was much the better one. Hence in January, 1895, the German Army High Command gave a trial order for 2000 Mauser rifles in caliber 7.9mm to shoot the same cartridge as then used in the Model 88 Rifle, stipulating that the rifles have a jacketed barrel. In October of the following year, a second trial order was placed for 2,085 rifles, this time of a 6mm caliber, minus the barrel jacket. However, on mature consideration, it was decided that the official 7.9mm rifle caliber was the correct one. In 1898, under the designation of "Infantry Rifle 98," the German Army adopted officially the 7.9mm Mauser Rifle without the barrel jacket.

In the following nine years, over 290,000 rifles were turned out at Oberndorf alone, while all the Government arsenals also participated in manufacturing this design for the complete re-equipment of the German forces.

This rifle, which formed the basis for much of the design of our own American Springfield Rifle, (on which a royalty of $200,000 was paid to Mauser by our

Army), with slight modifications, mostly of woodwork, sights, length and weight, remains to this day the official German Service Rifle; and again with only minor modifications of sights and finish is the basis for practically *all* the truly high powered sporting rifles of turning bolt design, with the sole exception of the Austrian Mannlicher type.

X

Mauser was among the first to appreciate the value of the self-loading principle. In 1896, he introduced his famous first Mauser Self-loading Pistol in caliber 7.63mm, a pistol so fundamentally right, that it and its cartridge are practically unchanged to this day. This arm, which Winston Churchill in his *History of the Sudan Campaign* credits with saving his life because of its rapidity of fire and its large magazine capacity, was a startling development in the Boer War. It is a remarkable achievement in that all the parts interlock and no screws or pins are used in the assembly of the weapon, with the exception of the screw which fastens the stocks to the arm.

Mauser demonstrated this new pistol to his Imperial Majesty Wilhelm the Second, at Katharinenholz range near Charlottenburg, on the 20th day of August, 1896. The Kaiser was extremely interested in this new arm, and did personal target shooting with it much to the delight of the inventor.

This weapon was really the first truly successful military automatic pistol. Up to the beginning of World War II, over 700,000 of these pistols had been sold. While it was never an official pistol of the German forces, it was a substitute-standard used during both World Wars by Germany. It is highly prized in the Orient and is to be encountered anywhere in the world. On August 20, 1896, during the course of the test by the Kaiser, Mauser promised to turn his energy to the development of a self-loading infantry rifle, promising the Kaiser to have such a development within five years.

With all the energy left, this untiring worker, whose sole thought was the expansion of German military might, devoted himself to producing semi-automatic arms. The range of his developments in automatic systems may be found by examining the contents of this book. Most of his rifles were either cumbrous or complicated. However, he did lay the basis for many designs, one of which turned up during World War II as the prototype of the essential locking system of the ultra simple German Gewehr 43, and of the famous Russian Degtyarov and German Model 42 Light Machine Guns.

XI

It is significant indeed, that with the passing of Peter Paul Mauser the only developments in his factory until World War II were a few elementary pocket pistols. The quality of his product was always kept up, but the true inventive genius of Waffenfabrik Mauser ceased with the Founder's death.

The very course of nations was made and altered by the inventions of Peter

Paul Mauser and by the manner in which they were merchandised. No corner of the earth was too remote, no nation too small to merit the attention of his organization.

Where Mauser rifles and pistols went to bolster the police and military authorities of a nation, there too went Mauser technicians and German service personnel to instruct—and to influence.

Paul Mauser died at the height of his success, an honored and respected citizen of his country. All that one man could do to further the ends of his country, Paul Mauser had done. He himself knew only the acclaim and military "glory" of the Germany in which he lived. It was left to his descendants to learn the utter, final futility of his efforts in relation to Germany's plans for world dominion.

XII

THE AMERICAN WHO CONTROLLED THE MAUSERS

Who was Samuel Norris? Just what part did he actually play in the development of the Mauser rifle? How did the Remington firm figure in Norris' association with the Mauser Brothers?

In contacts with the Mauser organization before the war, and in all my studies of original German contemporary records of the Mauser organization and its affiliates, I could never unearth any satisfactory answers to those questions.

Norris was European agent for Remington in the late 1860's. He entered into a contract to exploit the first Mauser rifle, then allowed his contract to go by default a bare year before the Prussian Government officially adopted the new weapon. Norris took out the first patent for the rifle in the United States . . Those were the essential facts derived from German sources. Paul Mauser himself never discussed his early dealings. The subject was one he disliked to think about as years brought him money and power and distinction. Why?

I could get no answer, either from the records or from European historians. A search of French, Austrian and Spanish records did nothing to explain the dearth of German information, but it did unearth many interesting sidelights on both Norris and the Remington organization of that period. As the pieces of the mosaic fitted together, they told an amazing story of adventure, business and enterprise in those dim early days of the breechloader.

The parts of that story which apply to the Mauser brothers directly has a tremendously important bearing on the contents of this book. Those parts, told largely in the actual written words of the incredible Samuel Norris himself, explain Paul Mauser's reticence to discuss the early days of his organization. It was a bitter page in his life, one he sought to forget.

The name Mauser is known wherever firearms are known. The name of Samuel Norris is all but unknown today. A powerful figure in his time, Norris once held the destinies of the Mausers in the palm of his hand. The contract he drew for the signature of the destitute brothers is one of the strangest documents in all the strange history of firearms.

"1. (The Mauser Brothers) inventors of a system of breech-loading rifle and central percussion cartridge, agree to sell, cede, and transfer to (Norris) the ownership of the said invention, with all rights which result therefrom in order to secure patents in all countries whatever; they engage likewise to transfer to him the ownership of every invention of this kind that they shall make hereafter and of every improvement that they shall bring to their system of rifle and cartridge."

That first article of the contract demonstrates graphically the extent to which Norris bound the impoverished Mauser brothers. For all this he was to pay them 80,000 francs over a period of fourteen years. If he personally decided to continue the contract that long! Even then there was a *hook* in the contract. Norris' contacts with the French were excellent, and he had sounded out the French military with a view to selling them the Mauser system to transform their Chassepot needle rifles to metallic cartridge arms. The French encouraged him to believe they might buy the system. Buried in the contract, therefore, we find the following:

"10. In any event, it is understood that in case Mr. Norris should cease to pay completely the annual sum stipulated above, he shall retain meanwhile as indemnity for his trouble the French patent."

XIII

During our Civil War the firm of E. Remington & Sons manufactured a variety of rifles for the North. Samuel Norris of Bristol, Rhode Island, sought War Department contracts as agent for Remington. In January 1865 he obtained one order for 5,000 carbines. These arms used the "split breech" mechanism as invented by Leonard Geiger and improved by Joseph Rider. When the hammer was thumbed back, the breechblock could be rolled back on its axis pin to expose the chamber for loading. The breechpiece (or block) was split. Several design improvements came later.

With the drying up of the American market for military arms, Remington sent Samuel Norris and his brother John to Europe to canvass new markets. Their success was phenomenal. No European nation had either our machinery or our knowledge. Then—even as now—war and the threat of war hung like a pall over the entire continent. Every nation felt itself menaced. All sought means of improved defense. The doors of all the chancelleries of Europe were ready to be opened by arms salesmen. And the brothers Norris were *super* salesmen. They walked with the mighty. They dined with kings. They became wealthy and powerful. And they had a strangle hold on the Mausers.

While Norris recognized the inherent value of the new Mauser rifle, his primary interest was in converting the Chassepot. He took Remington into partnership on the rifle deal knowing that they would bury the new bolt rifle in order to push their own rolling block rifle. Samuel Remington resolutely refused to face the fact that the bolt action was the coming military rifle; and the bankruptcy of his firm in 1885 stemmed in no small measure from that determination. It is to the credit of Samuel Norris that he tried to convince Remington that a change was coming.

When Norris was finally convinced that he had saturated the small European Nations with rolling-block Remingtons, that Samuel Remington would not push the bolt action, and that with the danger of approaching war France was against a change, Samuel Norris decided to save his yearly few thousand francs. He cut the Mausers adrift.

It is an irony of Fate that the one country where the Mauser never did make money was the one where Norris controlled the patents—France.

XIV

THE PERSONAL ACCOUNT OF SAMUEL NORRIS

In 1898, when the Mauser had proven itself a terrible weapon in the Boer War and in Cuba, when its name was a by-word in American newspapers, Samuel Norris sat down to write for posterity an account of his connection with the now famous rifle. The little Bristol *Phoenix* and the New York *Times* carried his account, an engrossing summary which is as remarkable for what it leaves out as for what it tells. Of the era in which the brothers Mauser worked (and nearly starved), Samuel Norris wrote:

"Every European Government was seeking a breech-loading system either as a new arm or as a transformation for muzzle-loaders. The English Government were about the first to decide, and they adopted the Snyder [Snider] as a transformation, really an American invention. Soon they began to transform their Enfields, the caliber of which was .57. [.577] Some years later the English Government adopted the 'Martini-Henry'—the name of 'Martini' applied to the breech mechanism, that of 'Henry' to the barrel. Again this system was mainly American, the invention of a Mr. Peabody of Boston.

"All the Continental Governments were alive to this important change of armament. The Germans had years before been the pioneers in breechloaders in their needle gun. Its caliber was .78. [15.43-mm or .601] In the base of the ball fulminate was placed, and the powder was held in a paper case. When the trigger was pulled the needle in the bolt shot forward, striking and igniting the fulminate, and the explosion followed. It had no effective gas check, hence the range was very small, and the gas came back into the face of the firer. However, the superiority of even this mechanism over muzzle-loaders was shown in the war between Prussia and Austria, and this hastened the efforts of Austria to get a breech-loading system.

"A grand commission was appointed in Vienna, the president of which was the Archduke William, cousin of the Emperor. The commission tested many systems, and decided to recommend to the Emperor the 'Remington.' His Majesty was invited to the arsenal to see the arm, and, as was expected, to approve of its adoption. He came with a staff of some seventy officers. After the inspection he was invited to fire the arm. I was present and remember well the brilliant gathering on the green in front of the targets. A young officer first fired the 'Remington' most satisfactorily, then the Emperor took the arm to fire. This arm and the cartridges had been made in Vienna to conform to the ideas of the commission as to caliber, form of bullet, and charge of powder. American metallic cartridge machinery was unknown in Austria at that time, hence the cartridges used for this trial, which were rim fire, were very imperfect. The very first one used by the Emperor failed to ignite; all others were successfully fired. This failure, with which the arm had nothing to do, proved fatal in Austria. All the newspapers attacked the Government for considering the arm, echoing the wishes of the hundreds in Vienna at that moment who were interested in other arms. It was even cabled to Providence that an accident had occurred to the Remington in the hands of the Emperor—a wicked misrepresentation. So fiercely was the Government assailed that the adoption of the Remington was abandoned, and trials of all systems were stopped. A few months later they took up the Wendel [Werndl] arm, an Austrian invention, destitute of merit, and adopted it. The Caliber was about .43. Still later that Government adopted the 'Mannlicher' system."

As Mr. Norris points out, the ammunition and the rifle were made in Austria and probably neither was up to the Remington-made standard. Certainly, no arm should be condemned because of the failure of a single cartridge to fire. To judge from the news accounts of the Vienna press, the system was rejected because a soldier was killed when a section of the split breechpiece was blown out. Indeed such a report was brought back to the United States by Lt. Col. C. B. Norton and figured in an official Congressional Report made on munitions at the Paris Universal Exhibition in 1867. Norton's report is confusing, since it carries a cut of the *improved* solid breechblock which was in use when he filed his report.

When Austria adopted the Werndl, it is noteworthy that the special cartridge designed for it had an *outside center fire primer* specially constructed to prevent escape of gas to the rear from the charge. The breechblock of the rifle itself was rotated up and to the left; and was so mounted and constructed that accidents such as that ascribed to the split-breech Remington could not conceivably happen. The danger of weak cartridge heads and blown out blocks appears to have been almost an obsession with the engineers at Steyr Armory, where the Werndl was designed.

In the United States there was so much dissatisfaction with the split breech mechanism that Remington's engineers worked to develop a more powerful, solid breechblock with a firing pin mounted in the block itself. This was the famous Remington Single Shot with which the Norris brothers blanketed much of the military world of that day. This new design was so staunch that even today specimens of the original rifles are in use in odd corners of the world—particularly in the Balkans and in South America; while in Sweden the design is currently manufactured as a shotgun lock mechanism. No stronger single shot rifle has ever been made.

The "improved" Remington was announced in the August 1866 issue of the U. S. Army and Navy *Journal,* but machine tool manufacture delayed actual production until the following Spring.

Indeed, in his own article Norris tells us indirectly of the new solid breech-block. He continues:

"Meantime the French Government were testing many systems. They inclined to the 'Remington,' and gave a small order of the dimensions they desired for exhaustive trial. At the moment this order was received in America an improvement had been made in that arm and much valuable time was lost in their delivery in Paris. Gen. Le Boeuf, President of the Committee of Artillery, was greatly annoyed at the delay, and as war clouds were gathering they hastily decided on the 'chassepot,' a bolt needle gun, using what was called a silk cartridge—that is, the case which contained the powder was made of silk, in the end of which was placed the fulminate and the needle by means of a spiral spring in the bolt when the trigger was pulled passed through and exploded it. For the purpose of preventing the escape of gas at the rear, what was called the *tete mobile* made of rubber was fixed on the end of this bolt. This entered the chamber, the explosion compressed it, and theoretically it was expected that it would prevent all gas and the debris from the burnt silk case coming out at the rear. This was not wholly realized. The debris did pass into and around the bolt, clogging the spring and the easy and proper movement of the bolt in the shoe in which it moved. After six or eight shots I have seen that the bolt could not be moved unless lubricated with water.

"The Greeks closely followed the investigations of the French at Vincennes, near Paris, and they decided on the Remington, and we made a contract for 15,000. These finally went to France during the Franco-Prussian War."

Greek Captain Alexandros Fountouclis, though unmentioned in the Norris narrative, was a prime mover in the Greek contract. He toured Europe testing rifle systems for the Hellenic Government and finally decided on the Remington. The placing of the order brought severe repercussions from Greece, and the original order was violently opposed in some circles.

The Norris narrative compresses into a few paragraphs a running fire account of the amazing range of travel and contacts of the next few years. The diligence, ingenuity and shrewdness of the Yankee brothers was perhaps without parallel in its day. In his own account, Norris constantly glides over the hardships and hairbreadth escapes. No one reading this later day account of his would ever fathom the strange depths of his Puritanic early-New England philosophy; no one could hope to catch even a glimer of the tortuous mental and ethical processes which made him and many of his Yankee peers wealthy and respected men. And no one could hope to assay the courage, the fortitude and the gruelling effort he poured into his activities.

His account continues:

"The Danes made exhaustive trials in Copenhagen, and decided on either the Remington or Peabody, and sent a commission to America to contract for one of those arms. My brother, Mr. John Norris, devoted himself for some time to this effort, and finally, the Minister of War, Gen. Rassloff, advised me to follow his commission to America, as the decision would be made there, but he would not give any assurance that the decision would await my arrival. However, I went. It resulted in a contract for 30,000 Remingtons, which was followed by other contracts. This was in 1867. Almost simultaneously my brother made a contract with the Swedish Government for a large quantity of Remington mechanisms, they proposing to complete the arm in Sweden. This they did.

"Meantime the Spanish Government had officers in America. The Remington was decided on by them for Cuba and orders were given. Then followed trials at Madrid, and the Remington became the adopted arm of that Government. I made three contracts in behalf of the Messrs. Remington, viz., for 10,000, 50,000 and 130,000. The Spanish Government had on its hands the war in Cuba and with the Carlists in Spain. It was an event to get either to or from Madrid, journeys which I made several times, and at considerable risk, when I passed through the Carlist lines having in my luggage abundant evidence of my dealings with the Spanish Government. However, I never had any serious trouble. The last order they tried to cut short by 30,000 arms, for they had more arms than were needed. When I heard of this, being in St. Petersburg, I went directly to Madrid and was most fairly treated by Gen. Jovillia, the Minister of War and the former Governor General of Cuba. He had to refer the matter to several commissions, and finally I got a favorable decision from the Council of State and these arms were delivered and paid for. In fact, while these contracts amounted to several millions of dollars, all was paid with a good degree of promptness. My relations with all Spanish officials were always pleasant. The caliber of the Spanish arms was .43.

"When the Viceroy came to Europe to invite the various crowned heads to the opening of the Suez Canal, I was requested to meet Ratib Pasha in London for the purpose of negotiating a contract for 'Remington' arms. It resulted in a contract being executed in the smoking room of Buckingham Palace for 60,000 arms. Several other contracts followed, all the guns ordered being manufactured at Ilion."

At this point, Author Norris in 1898 has set the stage to tell his story of his connection with the Mauser Brothers. Perhaps time has made him forget his dates: whatever the reason, he tells us that he first heard of the Mauser rifle in the seventies, while actually it was the Summer of 1867 when he first stumbled on his chance-of-a-life-time to dominate the small arms business of the World.

Norris continues:

"Early in the seventies I was in Vienna. My friend, Count Bylandt, afterward Marshal and Minister of War of the Austrian Empire, told me of a new arm which he had seen. All he knew about it was that it was the invention of two brothers, Mauser by name, and that it came to him from Stuttgart, Wurtemberg. It was shown to me. It was a bolt gun, and I saw in it features which I thought could be utilized in changing the Chassepot to a metallic cartridge gun. On my return to Paris I saw the Committee of Artillery, and found that if it could be done it would at least greatly interest them. I at once went to Stuttgart, and at the War Office found the whereabouts of the Mausers. Their home was in Oberndorf—some distance from a railway line. I went there, and found they were at work in the small Government armory in that place. Soon I was introduced to them. They looked liked crushed men—poor and working hard for their living. I found that the officials pooh-poohed their arm, and they had lost all hope. I asked them to my hotel, and soon I got the option to employ them and to exploit their invention if I should so elect within a certain number of days."

Thus the story was told for the edification of the world and for formal history by Samuel Norris in the year 1898. This is a *factual* story, and like most factual stories it is true—as far as it goes.

But for the *true* story we must turn to other records which show the mental processes of the famous salesman at the time of the happenings—the uninhibited records of a devoted husband to his loving wife, some long forgotten letters of Samuel Norris.

After receiving assurances from the French that they would be interested in a device to alter the Chassepot needle rifle to a metallic cartridge arm, Norris set out on his journey to hunt up the obscure German gunsmiths who had solved the problem.

His letter to his wife written on September 13, 1867 is a far cry from the stilted literary form of his later article in the Bristol *Phoenix*. Here we see the devoted husband, the shrewd psychologist, the sharp trader, the indomitable traveller, the ingenious operator, the utterly ruthless exploiter—all the complex characters which were rolled into one composite to produce a man who was the epitome of the Yankee international businessman of the period of the middle 60's. . . . No problem is ever approached directly, no cards are ever on the table. The inventors have an article of merit, but they are hopelessly poor? Then buy for a low price! You want their rifle system for a conversion? Then talk of cartridges, and of what might be done—at great expense—with foreign patents on the rifle. But don't mention France! Your employers, Remington, will not like your developing a competing rifle of more advanced design? They may cut you loose? Then dangle part of the new dirt-cheap contract before them! Suppose they don't exploit it properly, all you are concerned with is the Chassepot conversion!. . . .

But let us see the letter itself:

"September 13th, 1867.

"Hardly was a journey commenced with as much doubt as this was when I left Stuttgart and with such overwhelmingly satisfactory results. Suffering a good deal from a boil on my leg after five hours in the cars, I must say my heart almost failed me, when I came to get into the Post carriage at 9 A. M. to ride till one o'clock.

"However, I did it and was very comfortable. I got a good bed and this morning started out with my interpreter. I ought to say that when the landlord showed us to bed I asked him what sort of a town they had here. That was enough and he went on to talk of the gun factory and then about 'Mauser's wonderful gun.' Of course we were very ignorant. He offered to take us to the gun works and enable us to see this new gun. So you see the ice was broken at once. Well—so this morning we went and manifested great interest in the rude machines and work they were doing. Presently Mr. Mauser was brought forward and I told him I would go to his room and see his gun—so we went and talked and talked and examined and all the time I gradually led on without showing what my desire was. We talked much about the cartridges, he having a new one, then I said I had one—and asked the two Mauser brothers to come to the hotel in the P. M. to see it. We said good-bye and came off. At one, they came and never left till eight this evening. Well, I have an agreement *signed*—giving me the right to the whole invention—gun and cartridges for the world for 6000 francs per year for ten years. Cheap enough. I have all the details in black and white and well understood. Strange to say in the course of a talk with my excellent men—one of them said—"We have a brother in America working in Ilion.' A man I know very well and one of the best men in Remington's works. I said nothing, but when I found I was going to nail them I told them I knew him and saw him six weeks ago. They were surprised and delighted for they had not seen him for sixteen years. Besides they remembered that he had mentioned my name in his letters—(probably and naturally as Remington's representative in Europe). It gave them increased confidence, they said. Poor men, they are most intelligent, capable men, but have been kept down by every means possible and are now working twelve hours per day for four francs each. They have been told that their gun was the best before this and the Bavarian Government and I think it is the one these Governments will take.

"In fact—it is the cheapest thing yet produced and in many respects the best. I wanted it, as you know, for changing the Chassepot and they say they can do it. I hardly mentioned that however—though in the paper they agree to give me France in consideration of the patents I obtain for them in the next ninety days, if I don't carry out the whole agreement. So I have what I came for and more if I desire. I have not of course, determined what is best, but if E. R. and Sons do right, I think I shall let them in, in consideration of my continuing to act for them in the 'R'—This will induce them to do this, otherwise it might lead to a split, which I want to avoid. Besides, while I should be served, they will be too, for we need not force this, when there is any chance for the 'R'—then too, it's a good thing for them for it's the coming gun—and cheap as dirt and easily made, I mean quickly."

Norris knew the psychology of his employers quite as well as he knew the psychology of those he employed. For an interest in the new patent, Samuel

Remington was prepared to keep Norris in his employ: that way he could also control what might otherwise prove dangerous competition. And after all, the contract stipulated nothing about aggressively marketing the Mauser.

A closing note in the Norris letter of September 13th gives an additional sidelight on the strong tide of fortune which flowed for him in that year 1867:

"Well now I start at twelve midnight—ride in the Post carriage till four—then take the cars to Stuttgart arriving there at 9:30 tomorrow morning. Then at one, if I feel able and ride all tomorrow P. M. and night arriving Vienna, Sunday morning. Then I shall go to see Count Bylandt, he has one of the guns.

"It is certainly an incident this getting complete control of what promises to be an important gun in a little town in the interior of Germany through the tongue of a rather poor interpreter, for me and for them, as his German is quite different from theirs."

Norris continually worked at a feverish pace, his energy was all but boundless. The 25th of September 1867 found him in Liege, Belgium, where he wrote to his wife:

"My German boys have come and I have had them examining the Chassepot, and they say, without hesitation, they can easily change their arm as I want it changed. Certainly it promises well and I pray that it may result as well as we hope."

The following day he wrote his wife a graphic letter from the Hotel Suede in Liege:

"I am in my room, the two Mausers sitting at the table—my patent lawyer ditto at work at drawings. My interpreter examining the Chassepot—my cigar at my side —the table covered with all sorts of things—guns—cartridges—papers, etc. So much for the picture before me."

For once Samuel Norris was not telling his faithful wife the whole story—there was much more in "the picture before me."

There were two trusting, hard working men of genius blindly placing confidence and hope in him. To Samuel Norris they were merely "my German boys:" just two hard working, poverty stricken youths with a potential gold mine and no cash resources. For all his deep knowledge of how to influence people, Samuel Norris failed to plumb the depths or to assay the real stature of the simple men before him.

He could not look into the future and see sickly Wilhelm Mauser who, but a few years later, was to have opened for him all the Chancellery doors which by then would be forever closed to the once powerful Samuel Norris.

Nor could he look a few years still further into the future and see the brawny Paul Mauser who would talk with kings; and whose arms would sell in the tens of millions in all corners of the World.

If Samuel Norris had sensed those possibilities, if he had drawn a different contract, if he had pushed the new rifles as the inventors were led to believe he would—and as they themselves did—who can say how the history of the world might have been changed? Instead, he had drawn up in French— a language utterly unknown to the "parties of the second part"—the contract here translated:

"ME LOUIS DELBOUILLE
Notaire à Liège

Between *Samuel Norris,* manufacturer of arms, residing at Paris, rue de Berry, No. 2, ——— of the first part,

Wilhelm Mauser, armorer, residing at Oberndorf (Wurtenberg), and *Paul Mauser,* likewise armorer, and residing at Oberndorf, ——— of the second part,

The following agreement has been made:

1.—The parties of the second part, inventors of a system of breech-loading rifle and central percussion cartridge, agree to sell, cede, and transfer to the party of the first part the ownership of the said invention, with all the rights which result therefrom in order to secure patents in all countries whatever; they engage likewise to transfer to him the ownership of every invention of this kind that they shall make hereafter and of every improvement that they shall bring to their system of rifle and cartridge.

2.—The parties of the second part engage accordingly to do everything that shall depend on them to help Mr. Norris in obtaining patents for the said system in the different countries.

They give him for this purpose one or more irrevocable powers of attorney to take these patents either in his own name or in the names of the parties of the second part.

They will sign all papers and documents that shall be considered necessary and engage in general to do everything that shall be necessary to help Mr. Norris in the execution of his task.

3.—Mr. Norris shall have the right to dispose as he shall desire of the patents obtained whether under his name or those of the parties of the second part, who renounce taking out for rifles and cartridges of this kind any patents in any country unless as explained above by the mediation of the party of the first part.

The parties of the second part engage to sign and ratify all contracts of transfer of these patents to third parties if their intervention be demanded by the party of the first part; in one word, to perform all the acts necessary to transfer legally their ownership, the case arising.

4.—The parties of the second part undertake during six months, from the 13th of September last, to work for Mr. Norris or his representatives, if the party of the first part requires it, at the price of three to four florins per day for each of them, according to the expenses which they shall have to incur—

Mr. Norris will pay to them their fare if they be required by him to go to Vienna, or Paris or any other place.

5.—All the costs of patents are at the expense of the party of the first part.

6.—Besides the said party of the first part will pay to the parties of the second part within the period of ten years counting from this day for all the patents obtained and to be obtained the sum of 80,000 francs as indicated in the following articles, to wit:

Three thousand francs the first year	3000
Three thousand francs the second year	3000

Five thousand francs the third year	5000
Six thousand francs the fourth year	6000
Seven thousand francs the fifth year	7000
Nine thousand francs the sixth year	9000
Ten thousand francs the seventh year	10000
Eleven thousand francs the eighth year	11000
Twelve thousand francs the ninth year	12000
Fourteen thousand francs the tenth year	14000
Total: Eighty thousand francs	80000

This sum shall be payable by fourths at the expiration of each quarter.

7.—This sum of eighty thousand francs, likewise each annual sum, shall be distributed in the following manner among the different patents to be obtained:

A sixth for the English patent.

A sixth for the French patent.

A sixth for the U. S. patent.

A portion by equal parts for all the other patents.

8.—If from year to year, the number of these last patents obtained happen to increase there will be accomplished a new distribution proportional to the recompense due for each of them.

9.—In case Mr. Norris elects not to continue the payment of the recompense due for one of the patents obtained by him, after the manner of distribution indicated, the parties of the second part will have no other right than to take possession of this patent, without indemnity for Mr. Norris, who retains nevertheless the ownership of the other patents of which he shall have paid the recompense.

10.—In any event, it is understood that in case Mr. Norris should cease to pay completely the annual sum stipulated above, he shall retain meanwhile as indemnity for his trouble the French patent.

If however he has or should hereafter sell this French patent, he shall pay the sixth of the annual sum, as it is stipulated above.

If it should happen that the system Mauser be adopted at the same time by three of the four Governments, England, Austria, America or France, the total of the annual sum shall be paid to the parties of the second part.

11.—The sum of five hundred florins already paid to the parties of the second part by Mr. Norris will be entered on the account of the third annuity.

12.—The parties of the second part at all events shall have the right to receive and retain altogether the first two annuitis of three thousand francs each.

13.—The present arrangement shall become definite only if Mr. Norris gives notice of his acceptance within ninety days, dating from September 13 last.

14.—If the parties of the second part by their fault should cause delay in validating the patents, they shall be responsible to Mr. Norris.

Done in duplicate the 28th of September, one thousand eight hundred sixty-seven.

Approved, the preceding document except clause number 11 in which the words "on account of the third annuity" are replaced by the words "on account of the first annuity."

Signed { WILHELM MAUSER
PAUL MAUSER }

Besides it is expressly stipulated that the Messrs. Mauser can not under any pretext, directly or indirectly, transmit to third parties the rights which belong to the present contract of which the stipulations are applicable to the heirs of both parties.

Done in duplicate at the date above.

Signed { WILHELM MAUSER
PAUL MAUSER
SAMUEL NORRIS }

Witnesses:
EMILE DUPONE
WILLIAM SMITH."

The contract, a truly historic document in the field of firearms, tells its own story. The Mausers were to work for "three or four florins per day for each of them, according to the expenses which they shall have to incur." (A florin was worth something under fifty cents.)

For the first and second years they were to get 3000 francs per year. (The franc was worth about twenty cents then.)

The third year they were to get 5000 francs. (But by that time it was evident the new design could not be sold to France, so the contract was abrogated. Why waste 5000 francs?)

And as a crowning indignity, the final paragraph arranged to deduct from the first year's "annuity" of 6000 francs the sum of 500 florins "already paid." (Why let it go until the third year as originally stipulated? Who knows if the contract will be carried into the third year? Certainly a prophetic question. And another $250 of good American cash saved.)

Article 9 of the contract gave Mr. Norris an easy way out, once he was confident he could not swing a deal with the French Government. The Mausers had no recourse against him. All they could do was take control of their patent.

By 1870 he knew the French could not be sold, and rather than face the prospect of paying the Mausers the sum of 5000 francs—$1000—as stipulated for the third year, Samuel Norris exercised his rights under Article 9.

For a time the Mausers were stunned, but by then they had observed their employer long and closely, and they had learned much about how to proceed with War Ministries. The story of the official acceptance of the Mauser as the German service rifle has been told from the records of the Mauser organization in the earlier part of this essay.

When Samuel Norris failed to pay that pitiful third annuity, and lapsed his contract, he forfeited all rights to patents which would have made him far richer and more widely known than he ever dared to dream.

And now, what of the further association of the "parties of the first and second parts?" Let us again turn to the Bristol ***Phoenix*** account.

"My brother was interested with me, and later the Messrs. Remington became interested. We employed them about two years in working out the invention, making models, etc. Mr. Samuel Remington, who was in Europe, discouraged its presentation to military authorities, being anxious that the Remington should be the only arm to be energetically pushed. It was a grave error, for the inclination of military men was in favor of a bolt gun, following the German and French systems. The Mausers got discouraged and went home to Oberndorf, and soon after most wisely took their arm to Spandau, near Berlin, where all trials of arms and other military material were made. Soon the German Government adopted it, and it became the arm of the most powerful army on the Continent of Europe."

Thus wrote Samuel Norris in his quaint little New England home in the year 1898. What his true thoughts were we do not know—there are no late letters to show what he really felt. But he must have known that he had failed to take his tide at flood. His own "clever" business dealings had taken control of Mauser forever out of his grasp.

The original Mauser factory at Oberndorf.

JOHANN NIKOLAUS von DREYSE, 1798-1868.
Inventor of the needle gun, from which all modern bolt actions are derived. In France, where he worked for many years, his name is commonly written Jean Nicolas Dreyse.

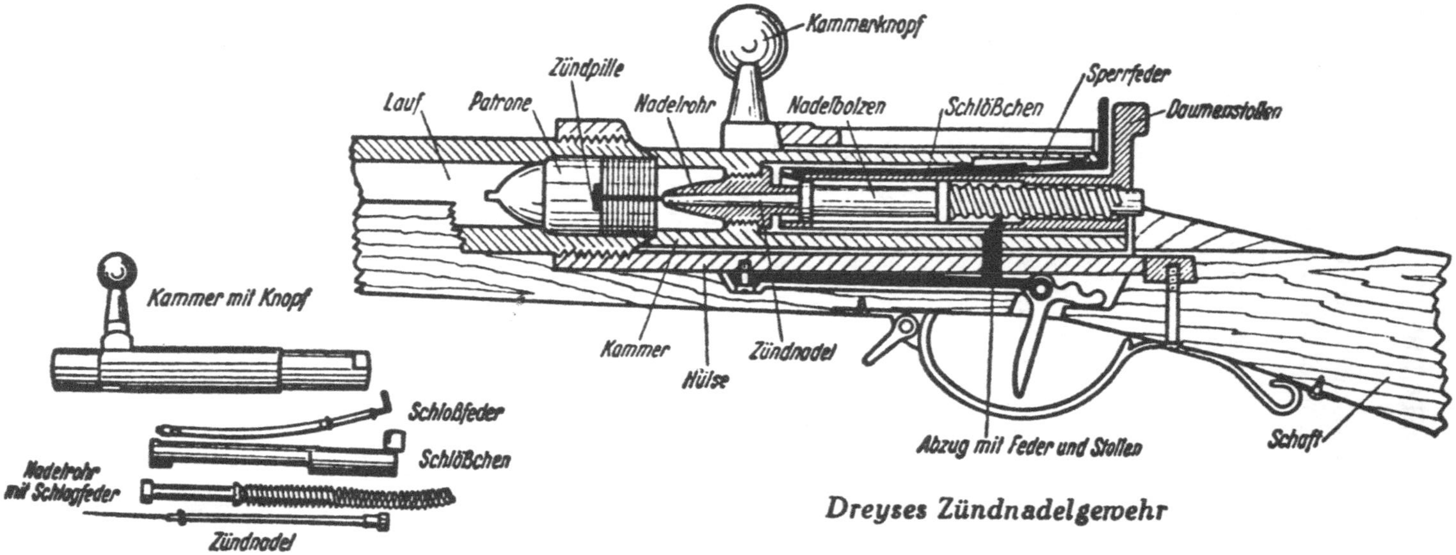

An original drawing of the Dreyse needle gun.

1. THE ZÜNDNADELGEWEHR OF NICOLAS DREYSE

The story of the Mauser rifles and of all military bolt action rifles begins with the development of the Dreyse Needle Gun, the famous Zündnadelgewehr.

The inventor of this first successful bolt action gun was Johann Nikolaus Dreyse, born in 1798 at Sommerda, near Erfurth, Germany. The firm bearing his name continued to manufacture firearms until the time of World War II.

Dreyse in 1809 worked with the already famous French gunsmith, Colonel Pauly, who did considerable experimental work on cartridges. At that time, it must be remembered, practically all rifles were loaded from the muzzle with powder and ball. Breech loading guns were dangerous because of gas leakage at the breech due to lack of a cartridge case which could act as a breech seal.

Dreyse built his first Needle Gun in 1827 and took out a patent on it in 1828. That patent covered both a spring needle and a fulminating cartridge, the gun itself being a muzzle loader. After many modifications, Dreyse produced his first unique bolt action rifle in the year 1838. This was the first production weapon ever developed in which the breech was closed by a sliding bolt operating on the principle of the common door bolt, in which a lever is raised to unlock the arm and the bolt drawn back in channels in the receiver to expose the breech for loading.

All earlier models of this arm had been muzzle loaders, the first experimental breech loading development appearing in the year 1836.

Many attempts had been made by Col. Pauly in Paris, among others, from 1808 to 1812 to develop a cartridge which would carry its own means of ignition within it. Indeed Pauly actually developed an experimental breech loader in which a detonating paper cap was attached to the paper cartridge and was fired by a needle. Dreyse doubtless drew on Pauly's experiments for his cartridge.

The Dreyse needle cartridge was a very peculiar one. It consisted of a lead bullet weighing 478 grains, having a length of 1.1 inches, and a diameter of .53 inch. Attached to the base of the bullet was a wad of papier maché in which was embedded some detonating powder, while *behind* this was the propelling charge of black powder varying from 66 to 75 grains retained in a paper bag.

When the bolt handle was lifted and the bolt drawn back, this peculiar cartridge was inserted in the firing chamber, the bolt was pushed forward, and the handle turned down to lock it. In most models a spring catch was provided at the rear of the bolt which had to be depressed and the striker pulled back by hand to cock, before the bolt handle could be turned up to permit the bolt to be withdrawn.

The rifle itself varied in calibre at different times of manufacture, ranging from about .60 to .66 inch. In all cases the bullet was smaller than the bore diameter and did not touch the sides of the barrel during its flight through it. Rotation was effected by the wad (or "sabot") at the base of the bullet which took the rifling and induced the twist. A typical example of this rifle, the 1860 model, has as 31-inch round barrel, rifled with 4-grooves. The twist is right, one turn in 28.82 inches. The caliber is 15.43 mm (.601)

The sliding breech bolt was hollow and housed a long needle about which was a spiral spring. This needle was the "striker."

When the trigger was pulled, the striker was released and driven forward by its spring, permitting the long needle point to pass through the paper powder container and through the powder itself until the point reached the *detonating powder* in the base of the bullet. The explosion of the detonating compound set off the powder, driving the bullet forward at a muzzle velocity of approximately one thousand feet per second. The powder was located *behind* the primer because Dreyse theorized that more complete powder combustion would result therefrom.

One of the serious defects of this rifle was the fact that the long needle was subjected to the fire and corrosion of the powder gases; hence its life was comparatively short.

Since this type of cartridge case did not seal the breech properly, flame would spit back around the bolt, and after the weapon had been used a number of times and was powder fouled, it was actually dangerous to fire it from the shoulder.

J. Scoffern in his "Projectile Weapons of War" (1859), and other contemporary writers and observers, tells us that in the early Prussian Wars the infantry quite regularly fired from the hip, because of the danger of back blast from the breech.

This rifle was officially adopted by the Prussians in 1840 and 1841 and proved extremely effective at that time because of its rapid fire by comparison with the muzzle loading weapons used by other armies.

The initial order from King William IV was for 60,000 Dreyse rifles. This continued as the *official* Prussian rifle until its replacement by the Mauser single shot metallic cartridge rifle in 1871.

With all its defects, the needle gun proved a most formidable arm. It was first used by the Prussians in their war against Denmark in 1864. When in 1866 it was used against the Austrian army which was armed with muzzle loaders, its success was so great that military authorities throughout the world realized that the day of the muzzle loading rifle was over. Wide experimentation was intensively started in Europe to develop adequate breech loaders.

The needle rifle was used by the Germans in 1870 against the French. The French opposed it with their Chassepot Needle Rifle, which was a decidedly superior arm. Its bolt head entered some distance into the firing chamber and had a steel shield, behind which was a thick wad of India rubber which helped to seal the breech against back flash when the weapon was fired. While this rubber plug or "obturator" was soon made too brittle by the heat to be completely effective, it still was such an advance over the Dreyse that the Germans utilized large numbers of captured Chassepots.

The tremendous value of the bolt principle of locking the breech was not grasped at that time. Indeed it was not until the American, James Paris Lee, developed his famous bolt action rifle in 1879 that the true worth of this system of operation in a military arm was fully appreciated except in Germany.

The defects of the needle gun due to the type of cartridge used was strongly impressed on the Germans by the course of the war of 1870. However, only the development of the metallic cartridge really opened the way to producing truly efficient breech loading arms. From this combination of Dreyse bolt action and the new metallic cartridge, in which the walls of the case expand at the instant of firing and act as a gas seal to prevent any gas from escaping to the rear, logically developed the next step. That was the first successful Mauser rifle, the almost unknown Mauser-Norris of 1867-69.

2. MAUSER-NORRIS 67-69. SINGLE-LOADER WITH TURNING-BOLT LOCK

This single-loader was developed during the years 1867-1869 by the brothers Paul and Wilhelm Mauser, with the financial support of an American, Samuel Norris.

At this period Lüttich was the fountainhead of European arms development, and it was there the actual manufacture was done.

No better description of this arm can be given—and certainly none more historically accurate—than that found in the records of the United States Patent Office, where this first successful Mauser rifle was originally patented.

AUS DEM AMERIKANISCHEN PATENT 78603 VOM. 2. JUNI 1868

(Originalabdruck der Einleitung und Zeichnungen)
Ferner patentiert in Frankreich, Belgien, England and diversen anderen Staaten

UNITED STATES PATENT OFFICE

Samuel Norris, of Springfield, Massachusetts, and Wilhelm Mauser and Paul Mauser, of Oberndorf, Wurtemburg, Assignors to Samuel Norris.

Letters Patent No. 78,603, dated June 2, 1868.

Improvement in Breech-Loading Fire-Arms.

The Schedule referred to in these Letters Patent and making part of the same.

TO ALL WHOM IT MAY CONCERN.

Be it known that we, Samuel Norris, of Springfield, Massachusetts, United States of America, at present residing in London, England, and Wilhelm Mauser and Paul Mauser, both of Oberndorf on the Neckar, in the Kingdom of Wurtemburg, have invented certain new and useful "Improvements in Breech-Loading Fire-Arms"; and we do hereby declare that

6 Sheets—Sheet 5.

S. NORRIS & W. & P. MAUSER.

Breech-Loading Fire-Arm.

No. 78,603. Patented June 2, 1868.

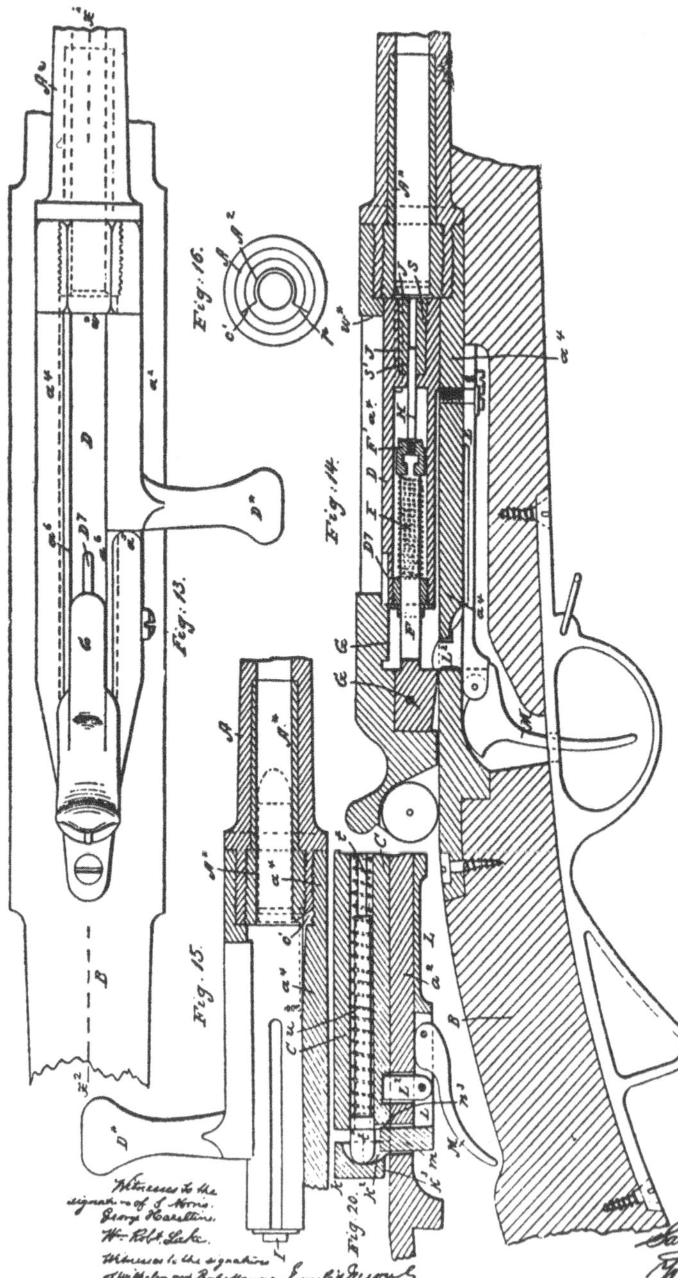

Witnesses to the signature of S. Norris.
George Haseltine.
Wm. Robt. Lake.

Witnesses to the signatures of Wilhelm and Paul Mauser.
Emile Wagner
William Smith.

Inventors.
Saml. Norris
Wilhelm Mauser
Paul Mauser.

the following is a full and exact description thereof, reference being had to the accompanying drawings forming a part of this specification.

The said improvements relate, first, to breech-loading mechanism of novel construction for fire-arms, whose breech is closed by a cylindrical block fitted to slide endwise in a chamber at the rear of the barrel. The said block is provided with a catch or projection extending from its surface, and, to close the breech, is turned upon its axis, so as to bring the said catch in front of a shoulder on the side of the breech-chamber, and thereby keep the said block securely up to the breech. In opening the breech, the catch is turned back into line with a longitudinal aperture, which allows the said catch to pass, and the block to be drawn back.

The cartridges are ignited by the blow of a firing-pin, which passes through the cylindrical block, and is driven forward by a spring, which is held by a catch, and is released by means of an ordinary trigger.

The said improvements consist chiefly in the peculiar construction and arrangement of the main-spring, and in the devices employed in connection with the trigger for holding and releasing the said spring. The latter is attached at one end to the extremity of a handle, which projects radially from the rear of the cylindrical breech-block, the other end of the said spring being free, and arranged exactly opposite the centre of the said block, which is in line with the barrel.

The rear end of the firing-pin lies in contact with a head formed on the end of the spring, and when the latter is released from its catch, it drives the said pin forward into contact with the cartridge in the charge-chamber. The said pin is kept in contact with the head of the main-spring by a light spiral spring, which bears against the rear end of the pin, and the forward end of the breech-block.

The head of the main-spring lies close to the end of the cylindrical block, except when the piece is cocked, and is formed with an inclined surface or cam. This surface is arranged to act in such a manner upon the catch or stop that when the cylindrical block is turned in the proper direction upon its axis, the main-spring is forced back behind the said catch, the piece being then cocked. This movable stop is fixed upon the end of an elastic bar, which is secured at its forward end to the under side of the metal shoe, wherein the breech-chamber is formed. This bar is so formed and arranged that, when free, its elasticity forces it upward, and keeps the catch in its proper position to hold the main-spring back.

A central longitudical aperture is formed in the elastic bar to receive a small arm, which is pivoted in this aperture near the rear end of the said bar. The free end of this arm lies in contact with the under surface of the

---

ONE OF THE 6 SHEETS OF PATENT PAPERS SIGNED BY NORRIS AND THE MAUSER BROTHERS, NOW IN THE UNITED STATES PATENT OFFICE. THIS WAS THE FIRST MAUSER PATENT EVER FILED ANYWHERE.

breech-shoe. A small lever, which is also pivoted to the elastic bar, extends under the free end of this arm, and lies in contact with a shoulder or projection on the trigger, which is arranged in the usual position at the under side of the stock. When this trigger is pulled, the said shoulder or projection forces the small lever upward, and the said lever acts on the small arm, which, being thereby pressed against the bottom of the breech-shoe, forces the elastic bar down, and draws the catch away clear of the spring. The latter being then released, flies forward, driving the firing-pin sharply against the cartridge, and causing its explosion.

The breech-block is kept from being accidentally turned into such a position as will allow it to escape from its chamber by a catch connected with the elastic bar. When the block is turned with its stop in front of the shoulder which keeps it up to the breech, this catch fits into a recess formed on the surface of the head of the mainspring, and prevents the turning of the block till the trigger is pulled.

The forward portion of the block may be made separate from the main or rear portion, and attached thereto in such a manner that the rear portion will turn on its axis without turning the forward portion, but so that both parts are kept firmly together endwise. The firing-pin extends through the central perforation in the said block.

This arrangement of parts is especially applicable to central-fire cartridges, but our improvements may be adapted to rim-fire cartridges by a simple modification of the end of the firing-pin and block.

The said improvements also consist in the construction and arrangement of the devices for removing the shells of the exploded cartridges from the piece. The said shells are drawn from the charge-chamber by an extractor, which is attached to the breech-block. This extractor is elastic, and has a claw or hook, which, as the breech is closed, passes over the rim of the cartridge, into the proper position to take hold of the same, and draw it back when the breech is opened.

The end of the barrel or charge-chamber is chamfered or bevelled away, for a portion of its circumference, to receive the hook or claw of the extractor, which is thus kept in the proper position to extract the cartridge, without interfering with the turning of the block. This form of block with the chamfered end of the barrel is more especially applicable when cartridges are used which have a solid metallic disk or flange at the base.

The proper action of the block and extractor may be also effected by attaching a loose piece to the end of the block to carry the extractor, which lies in a groove or guide-way formed in the side of the breech-chamber, and prevents the turning of the loose piece with the rear portion of the said block, when the breech is being opened or closed.

This last-described arrangement is preferable with ordinary metallic cartridges.

An aperture or slot is formed through the bottom of the breech-chamber, and in this aperture is arranged a lever, which forms the device for ejecting the cartridge from the said chamber. The extractor draws the cartridge back till its rear end rests on a nose or propection on the extremity of the long arm of this lever. When the piece is closed, this projection lies a little below the bottom of the chamber so as not to interfere with the movements of the block. The short arm of this lever extends into a longitudinal groove or channel, formed along the breech-block to allow the same to pass over the said projection. This groove is formed with a lateral extension in the proper position to allow the block to clear the ejector when turned on its axis.

The longitudinal groove extends nearly to the forward end of the block, and the shoulder formed by its termination strikes the short arm of the ejector, when the block is drawn back, and jerks the long arm sharply upward, thereby throwing the cartridge out of the piece.

The said improvements relate, secondly, to certain modifications or alterations in needle-guns, and more especially to the arm known as the "Chassepot" gun. In the said "Chassepot" gun, the breech is closed by a cylindrical block, which is provided with a metal needle-guide, and a disk or ring of prepared India rubber, or like substance, to prevent the escape of gas at the time of explosion. This gun is constructed only for firing cartridges with cases of paper or other soft material. These cases being consumed, no extracting or ejecting-devices are required. The said cartridges are ignited by a needle, which is driven forward by a spiral spring placed in the centre of the cylindrical block.

The chief object of this part of our improvements is to adapt the said "Chassepot" gun to the firing of metallic cartridges. For this purpose, we remove the India-rubber washer and the metal guide from the breech-block, and attach to the forward end of the said block our improved extracting-device, as described in a former part of this specification. We also form an aperture through the bottom of the breech-chamber to receive our improved ejecting-device, which may be, in this case, attached to the under side of the breech-chamber or to the elastic bar below the same. If desired, this ejector may be dispensed with, and the cartridges thrown out by simply reversing the arm.

We remove the needle, and substitute for the same a firing-pin, which is arranged to strike the rear of the cartridge when forced forward by the spiral spring, which we retain.

The end of the barrel which is screwed into the breech-shoe is chambered for some distance, to receive the end of the block; or a ring or bush may be fitted into the end of the barrel, and a chamber so formed in the said bush that the rim of the cartridge lies flush or nearly flush with the rear of the barrel.

The breech-block is adjusted by means of a handle, which forms the stop for holding the said block when the breech is closed.

3. THE EARLIEST MAUSER RIFLES

In the interests of historical documentation, the following literal translation from the German of the description given by Theodor Schmid, Director of the Waffenfabrik Mauser A.-G. at the turn of the century, is of distinct value.

Schmid was a personal friend and confidant of Paul Mauser, and his description of the earliest Mauser rifles, as well as his sidelights on Norris, had the approval of the great inventor.

These details, therefore, constitute the best available authority on one phase of Mauser's activities and weapons which has long been the subject of conjecture and misinformation. (Note: Schmid uses the term "chamber" in reference to the long cylindrical bolt body. The term "shoe" means the receiver.

Single-Loader With Cylinder Lock, Mauser-Norris C. 67/69

History: The single-loader described below was created in the years 1867-1869, with the financial support of the American Samuel Norris, whose assistance enabled the brothers Paul and Wilhelm Mauser to work out in Lüttich the novel ideas, conceived especially by Paul Mauser, relating to the improvement of the then still very imperfect breech-loading system with a cylinder lock, with particular consideration to the use of metallic cartridges. The efforts of the two brothers were crowned by success; the experimental weapons then produced in Lüttich (see the pictures which follow for a complete representation of these rifles) embody already all the essential improvements which carried the later Mauser rifles, first as single-loaders, to triumph and glory all over the world.

Principle: As the picture shows, this first complete Mauser-Norris rifle already displays the following arrangements, so important in principle:

1) The lock is designed for self-cocking.

2) The cylindrical block, the so-called "chamber," bears on its forward end a rotatory bolt-head which, due to its rotatory connection with the chamber is not affected by the rotation of the latter, so that when the chamber is opened or closed, this bolt-head retains its position, *i.e.* does not turn.

(This arrangement resulted in a better obturation of the then still used paper cartridges in particular, inasmuch as due to the nonturning of the bolt-head during the closing, the bottom of the cartridge was

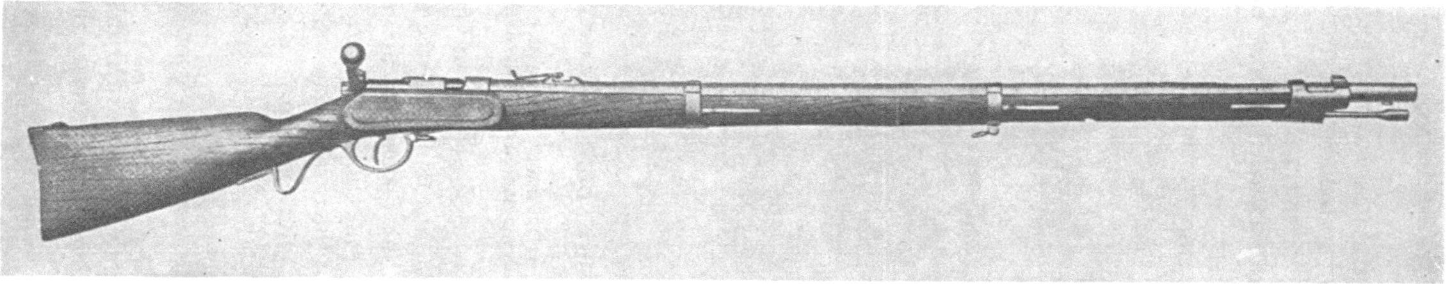

MAUSER-NORRIS 67-69

Right side view of the original Mauser, the "Mauser-Norris," with action closed. This rifle, first patented in the U. S. jointly by Paul Mauser and Wilhelm Mauser, inventors, and Samuel Norris of Springfield, Mass., as financial backer, is the parent arm from which the long line of Mauser rifles descended. This first design incorporated many of the basic Mauser charactertistics still in use in all bolt action military rifles.

Illustrations from original early Mauser records.

not exposed to friction, and therefore remained undamaged. From the point of view of the planned introduction of metallic cartridges, however, the rotatory bolt-head represented the great advantage of permitting the use of an extractor fastened immovably to the bolt-head.)

3) During opening, the fired empty cartridge is pulled back by a curve on the breech-shoe.

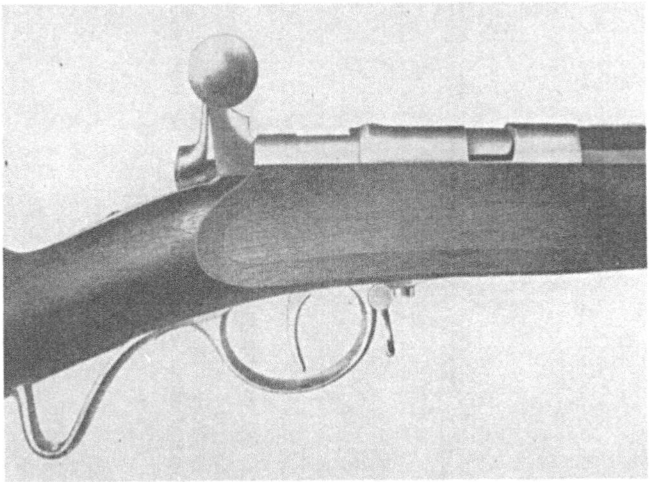

MAUSER-NORRIS 67-69 RIGHT SIDE VIEW OF RECEIVER SECTION

Close-up view of the first Mauser bolt action metallic cartridge rifle. Note unusual placement of bolt handle.

4) During opening, the striker is pulled back so far as to make its tip move behind the front surface of the breech face of the chamber, respectively of the bolt-head, thus preventing the possibility of a premature ignition due to the projection of the tip of the striker during loading (forward drive of the cartridge into the barrel).

5) The arrangement includes the extractor, fastened stationarily to the bolt-head (as mentioned under 2), which is especially valuable when metallic cartridges are used.

6) A special ejector is provided, which, too, is of great importance for the use of metallic cartridges.

1st Model: The following are the *details* of the mechanism of the Mauser-Norris rifle designed along the general principles outlined above:

The *breech shoe* is perforated, and on its rear face, behind the perforation on the left-hand part of the shoe, there is a curve, along the surface of which the handle is driven a little backward when the cylinder (chamber) is turned up, thereby forcibly dislodging the fired cartridge from its position.

The *bolt-head* is seated on the forward end of the chamber, and is protected by a combination of projections and recesses against longitudinal dislodgements.

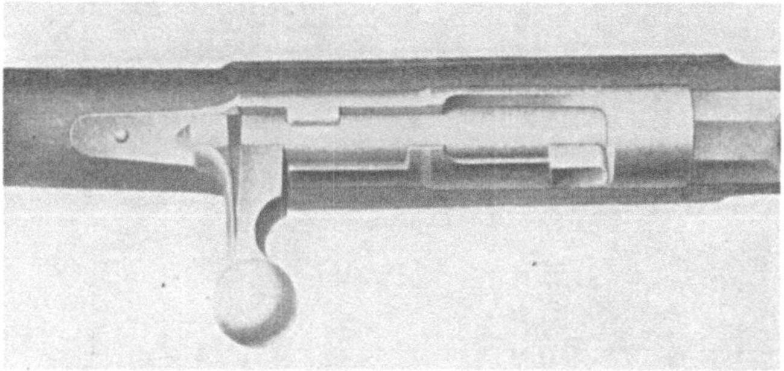

MAUSER-NORRIS 67-69. TOP RECEIVER VIEW
The locking lug on this arm is an exterior one locking down in the receiver on the right side behind the chamber section. This was the first crude form of the now world famous Mauser bolt.

In order to produce a *double-side locking device,* there is a comparatively big stud on the chamber, below the forward end, in the same plane with the handle; a matching counter-stud is provided on the opposite side, but this counter-stud is located slightly farther forward, and is completed by a stud located on the bolt-head and extending in the same direction.

The formerly used thin firing needle has been replaced by a *strong* striker. The mainspring, fashioned as a leaf-spring, is attached to the handle above, and forms one piece with the snap, the lower end of the spring being fashioned so as to form a cup to receive the striker. A spiral spring is pushed onto the forward part of the striker; this spring tends to keep its rear end pressed into the cavity of the cup-shaped snap, so that when the snap is moved backward during the cocking of the mainspring, also the striker will follow this backward movement.

During closing, the *self-cocking of the lock* is effected as follows:

The rear end of the chamber has a recess, into which the upward-pointing sear snaps when the chamber is moving forward. Thus the sear stops in front of the snap, the inside face of which has a curve-shaped chamfering. When the cylindrical block is turned closed, this curved chamfering slides on the cocking stop, whereby the snap is forced back, and the mainspring is cocked.

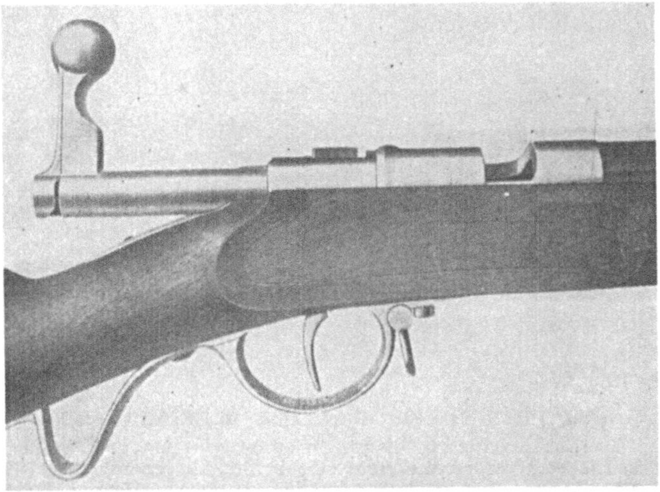

MAUSER-NORRIS 67-69. RIGHT SIDE VIEW OF RECEIVER SECTION WITH BOLT TURNED UP AND DRAWN BACK TO OPEN ACTION FOR LOADING

In this type of bolt design the bolt way is so constructed that the cartridge was dropped in through the top of the action, and the ejection was also from the top. It was not necessary to thrust the cartridge directly into the chamber by hand as in other designs.

During opening, the striker is pulled back as follows: Matching shoulders or chamferings are provided on the forward part of the striker and on the rear end of the stock of the bolt-head, and when the chamber is turned up, their concerted action impart a slight backward movement to the striker. The striker can occupy the position of ignition only when the locking is completed, *i.e.* the chamber is turned closed, when the shoulders and chamferings face each other in such a manner as to leave the striker free play to protrude sufficiently over the frontal face of the bolt-head.

The *trigger mechanism* shows a trigger rod with the sear, which latter also serves as cocking stop. The sear is attached to the rear end of the rod in such a manner that when the trigger is pressed, it is forced down, so that it is separated from the snap, whereby the action of the mainspring causes the snap to fly forward, carrying along the striker which lies in its cavity.

The *extractor* is attached laterally to the bolt-head; a special groove is provided for it in the inside wall of the breech shoe, which serves as its guide-way when the chamber is opened and closed. When the chamber is moving back and forth, the extractor slides in this groove, whereby also a rotation of the bolt-head is prevented during the turning of the chamber up and down.

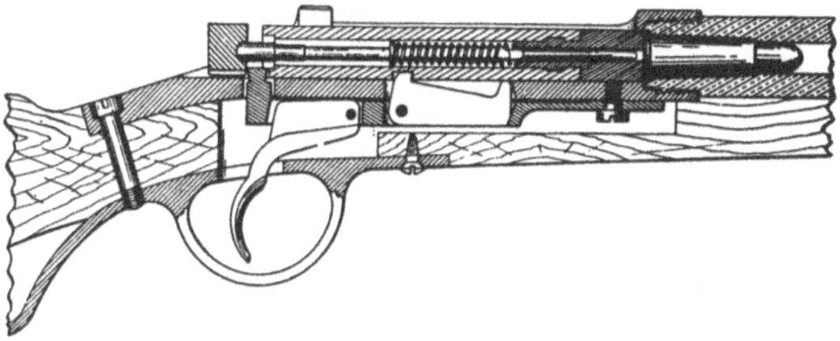

MAUSER-NORRIS 67-69 RIGHT SIDE SECTIONAL VIEW OF RECEIVER WITH ACTION CLOSED

The rifle is cocked and loaded ready for firing by pressure on the trigger. Note the shape and position of the powerful ejector whose rear end is cammed down as the bolt is pulled to the rear, thereby lifting the front point of the ejector sharply to knock the empty case out the top of the rifle.

The *extractor* is lodged in the bottom of the lock shoe, in such a manner as to permit it to swing. In the chamber there is a recess to receive the rear end of the extractor (bent as a hook in front) when the locking is effected, so that the extractor is kept motionless. When the chamber is pulled back, the rear end of the extractor slides out of the recess, so that it is swung into its action position with its hook or ejection shoulder upward, thus giving the extracted empty cartridge case the necessary impetus for the ejection.

Patents: Of all the patents taken out for the innovations described above, United States Patent No. 78603 is of the greatest interest to us right now, for this single patent, awarded in the name of Norris and the Mauser brothers, is actually a summary of all the innovations covered by the various individual patents issued in other countries. We reproduce here literally its introduction and claims, as well as the drawings which form part and parcel of it. As regards the patents awarded in other countries, not requiring an identification of the inventor, we quote here abstracts from Austrian Patents 17864 XIX[9] and XIX[26] which are in German, and will therefore be clear to the reader. These patents, representing the fruit of the work of the Mauser brothers (in particular of the practical labor of Paul Mauser), show beyond the shadow of a doubt that the weapon covered by these patents, and which became known to the general public as the "Mauser-Norris rifle," or simply as the "Norris rifle," actually embodies the principal technical features of the Mauser lock, wherefore it is of utmost importance not only from the historical point of view, but also practically in the development of Mauser rifles, and of portable firearms in general.

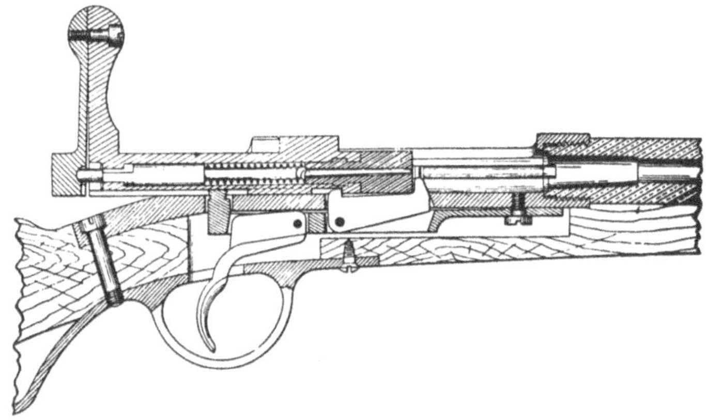

MAUSER-NORRIS 67-69. SECTIONAL DRAWING OF RECEIVER WITH ACTION OPEN SHOWING ALL DETAILS OF CONSTRUCTION

Note particularly position of ejector and details of spring support at rear of bolt handle. On forward movement the bolt rides the ejector down below bolt level.

2nd Model: Some time after the completion of the weapon known as "Mauser-Norris rifle," another model rifle was perfected; as regards its lock arrangement, this later model is identical with the single-loader described above, but it embodies certain detail improvements, and it represents in fact the intermediate stage between the "Mauser-Norris rifle" and the German M. 71 rifle. As far as it can still be ascertained, this rifle shows the following changes:

The *mainspring* (attached to the handle in the previous model) is dispensed with, and replaced by the strong spiral spring, lodged in the chamber and surrounding the striker.

A guide-rib is provided for the *chamber;* this guide-rib is joined to the foot of the handle, and it forms the breech base. Its forward part, protruding over the end of the chamber, has a transverse groove or notch, and when the breech mechanism is opened, this groove or notch extends with a shoulder over a nose of the bolt-head (which is inserted into the chamber end without any coupling, *i.e.* without any stop to prevent its longitudinal displacement, although it does not share the turning movement of the chamber), carrying the bolt-head along when the chamber is pulled back.

The *snap* is pushed on the rear end of the striker, as a separate part, and is kept in its place by the separate striker nut, screwed onto the striker behind the snap.

The *self-cocking* is effected as follows: A projection is provided on the forward end of the snap, and a chamfering or recess on the rear end of the chamber, so that they engage each other when the locking is effected. When the chamber is turned up, the projection and the recess slide along each other, whereby the snap is forced back, and the spiral spring, surrounding the striker and serving as mainspring, is cocked.

Furthermore, this rifle was the first to have a *safety device* of the cocked lock, in connection with the snap; this safety device consists of an adjustable retaining piece which takes up a position between the chamber and the snap when the handle is pointing upward, thus preventing the snap from flying forward.

Patents: The financial backer who had applied for a patent for the first "Mauser-Norris design" for his own account and in his own name, neglected to have the above improvements patented. The Mauser brothers were not in the position to apply for the necessary patents, for at that time whatever modest financial resources they possessed were needed primarily for the expenses of the production of the various test and trial rifles. Moreover, it was anyway impossible to acquire a sufficiently ample patent protection in the German States, for it is a well known historical fact that in the absence of a universal German Reich patent law, there was no patent protection available in those days for the entire territory of the Reich."

4. THE FIRST PRODUCTION MAUSER RIFLE, MODEL 71. GERMAN INFANTRY WEAPON

As already pointed out in the Historical Foreword, this arm while officially adopted in 1871 was actually issued in February 1872, with the improved safety demanded by the Testing Commission.

This arm was a single shot breech loader utilizing the bolt principle of the Dreyse and resembling it in many exterior details. However it employed a new cartridge whose case was drawn brass. This cartridge measured 3.07 inches over all and weighed 660.5 grains. The bullet was lead; its length was 1.0827 inches, its diameter (less patch) was 11 mm, (.433 inch), and its weight was 386 grains. The initial charge was approximately 77.16 grains of black powder which developed a muzzle velocity of about 1425 feet per second and gave a maximum range of about 3200 yards.

The rifle itself weighed 10 pounds 4 ounces (1 pound 8 ounces more with bayonet). The overall length was 4 feet, 4.75 inches (6 feet .5-inch with bayonet). The barrel was 33.46 inches long and was rifled with 4-grooves of .0157 inch depth. The rifling made one turn in 21.65 inches (50 calibers) to the right. The sight adjustment ran from 300 meters (328 yards) to 1600 meters (1750 yards). The bore diameter was nominally 11 mm (.433).

This rifle was such a revolutionary development in the art of war that a fairly complete description of it is warranted here. Furthermore, millions of these rifles were distributed throughout the world, very large numbers having been sold in the U. S., particularly by the firm of Bannerman of N. Y. C. Moreover, they appeared in some quantity in use even in World War II in Europe. Ammunition for them was generally manufactured in Europe as late as 1937. Large numbers of them were adapted first in the late '90's and later in the early '20's to handling shot gun shells. These arms therefore may be encountered in the U. S. *in general use* as well as in arms collections.

The illustrations here used to show the functioning of this weapon are taken from contemporary drawings officially made by Mauser.

The bolt is fitted into the receiver of the rifle to permit it to travel backward and forward as the bolt handle is raised or lowered and the bolt drawn back or pushed forward.

The bolt cylinder is perforated and has the striker surrounded by

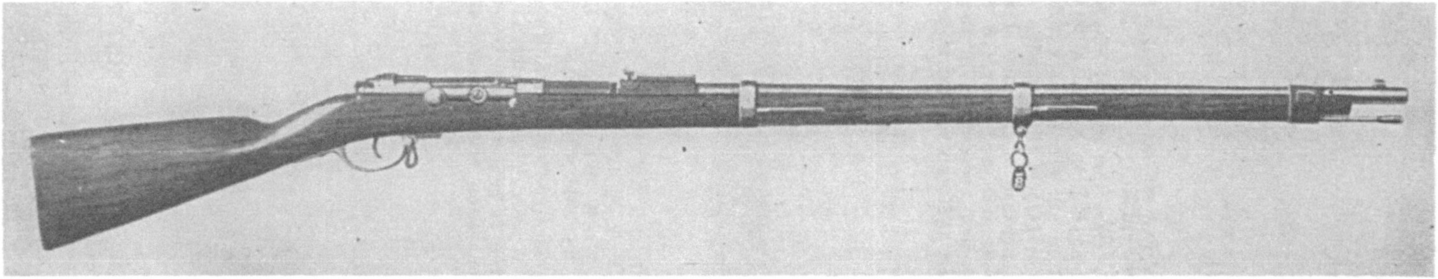

MODEL 71 GERMAN INFANTRY RIFLE

Right side view with action closed showing original Mauser single shot metallic cartridge rifle adopted by the German Army in 1871 and actually introduced into service in the following year.
This model used the removable bolt head and split bridge receiver later used on turn-bolt Mannlicher rifles.
The German Gew. 1888 bolt and receiver were evolved from the M. 71, not from Mannlicher designs.
Model 71 rifles, both original and converted types, were used in the Balkans even in World War II.

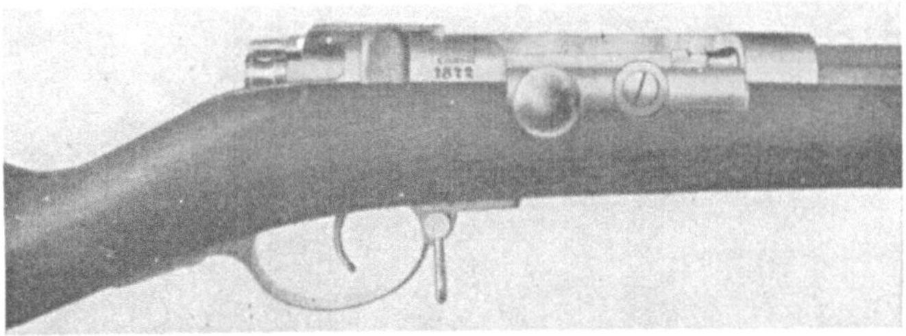

MODEL 71. RIGHT SIDE VIEW OF RECEIVER SECTION ACTION CLOSED

This design differs radically from the current Mauser. The bolt is of entirely different construction but incorporates the basic cocking features found in all later Mauser rifles. This specimen was manufactured in 1872 at the Spandau Arsenal. While called the 1871 Model because of its original adoption in that year, actual specimens were not delivered for field use until 1872, after changes had been made in the safety. The safety system is essentially the same as that in use today.

MODEL 71. RIGHT SIDE VIEW WITH BOLT UP AND TURNED BACK TO OPEN ACTION

Construction of cocking-piece differs from the earlier Mauser-Norris. This rifle is usually encountered stamped with the year of manufacture. It is the really direct parent in cocking, extraction and operation of all later bolt action Mausers.

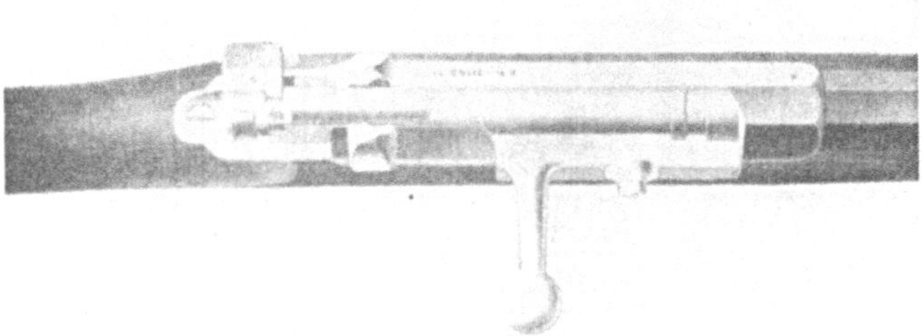

MODEL 71. TOP VIEW OF RECEIVER SECTION WITH ACTION CLOSED

Note type of bolt safety which very closely resembles that of the latest type Mauser. This system of safety has never been substantially improved upon.

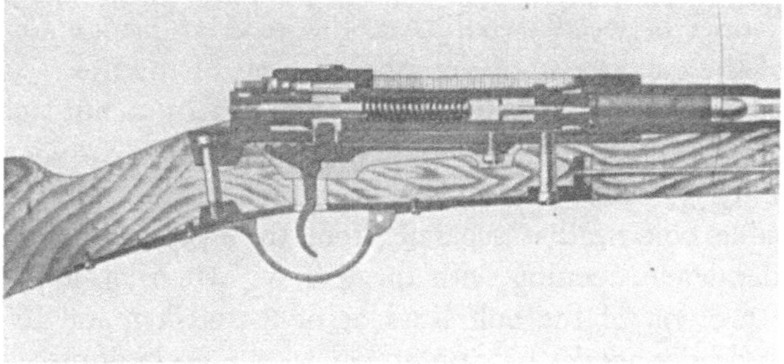

MODEL 71. RIGHT SIDE PHANTOM VIEW SHOWING RIFLE COCKED AND LOADED READY FOR FIRING

The simplicity of this first design has never been surpassed. The bolt cylinder is pierced to receive the striker pin and spring. The bolt handle or lever is part of the cylinder. The guide block on the upper surface of the cylinder working between the lips cut into the receiver, prevents rotation of the cylinder until the guide is clear of the slot. The striker in this design is permanently attached by a nut and screw to the cocking-piece. The cocking-piece can move only backward and forward and *cannot* rotate. This is provided for by having the striker elliptical in section, its shape changing at different sections of its length.

The bolt head is removable and can be readily detached from the bolt cylinder. However, since it rotates with the cylinder when the action is opened or closed, the position of the striker and spring are not affected. (This two-piece bolt is cheaper to manufacture than the stronger one-piece type found in all Mausers of modern design).

The cam slot in the rear section of the bolt cylinder acts on the cocking-piece as the bolt handle is lifted to withdraw the striker pin within the cylinder of the breech bolt.

The extractor works in a groove in the left side of the receiver and in the movable head of the breechblock (or bolt head).

The ejector is mounted in the trigger spring and works through the receiver bottom.

The safety catch on the cocking-piece works upon a spring. When applied it drops into a notch provided for in the bolt and prevents the striker from reaching the cartridge.

When the bolt handle is turned up to the left as far as it will go, the head of the breechblock and the cocking-piece remain in the same relative position. The cam at the rear of the cylinder forces back the cam of the cocking-piece to partly cock the striker.

The bolt is then drawn back to extract and eject the cartridge or empty case in the firing chamber.

The ejector moves in a groove in the underside of the bolt cylinder, hence its forward claw projects only when the bolt is in full rear position.

its spring mounted within it. Raising the bolt lever permits the bolt to be drawn back. A guide block on the upper surface of the bolt cylinder works between two lips in the receiver bridge to prevent rotation of the cylinder until the guide is clear of the lips.

The striker is not removable and is attached by a nut and screw to the cocking piece. The cocking piece can move backward and forward only. It cannot rotate, as the striker is elliptical in section. The movable bolt head is separate from the cylinder; and rotating the cylinder when opening with the lever or when closing does not affect the position of the bolt head or of the striker and its spring. The bolt cylinder works by cam action on the cocking-piece to cock the striker as the bolt handle is manipulated.

MODEL 71. RIGHT SIDE PHANTOM VIEW WITH ACTION OPEN READY FOR INSERTION OF CARTRIDGE

Knob of bolt handle is not shown in drawing.

Note details of cocking cam surface outlined in the cocking-piece. Unlike modern Mausers, these early types had a removable bolt head. It will be observed that the striker pin is back inside the bolt cylinder but is not yet held at full cock.

The extractor works in a groove in the left side of the bolt cylinder, and is of spring steel. The ejector is mounted in the trigger spring and works through the bolt cylinder and the receiver.

A safety lock fitted to the cocking-piece works upon a spring. When applied it drops into a notch in the breechblock (or bolt) and prevents the striker from reaching the cartridge.

Operation

The bolt handle is raised and the bolt is drawn back as far as it will go. A cartridge is inserted in the firing chamber. The breech block is pushed forward until its guide is clear of the receiver bridge slot. It is then turned down to the right to close and seal the breech;

and since the nose of the cocking-piece bears against the cylinder cam, it also completes compression of the striker spring ready for firing. Pressure on the trigger is transmitted to release the striker. The striker spring drives the striker forward to fire the cartridge.

When the bolt lever is turned up to open the breech, the head of the bolt and the cocking piece remain in the same relative positions as already indicated. The cam face of the bolt cylinder forces back the cam on the cocking-piece, thereby drawing the striker back into the bolt.

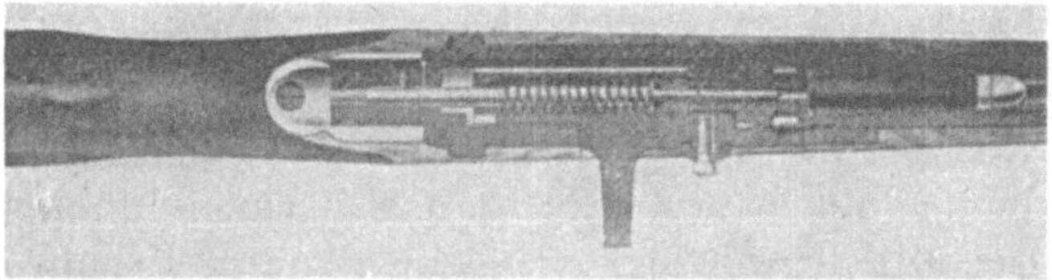

MODEL 71. TOP PHANTOM VIEW SHOWING DETAILS OF RIFLE WITH CARTRIDGE IN CHAMBER AND WEAPON READY TO FIRE

Knob of bolt handle is not shown.

After insertion of a cartridge in the feed way, the bolt handle was pushed forward to chamber the cartridge and permit the extractor to snap over the rim of the case.

The bolt handle was then turned down to the right. Since the nose of the cocking-piece was bearing against the bolt cylinder, turning down the bolt handle completed cocking the arm and compressing the striker spring by direct rearward thrust through its cam action.

Note that the shock of discharge is taken principally by the rear locking surface behind the bolt handle.

The bolt handle locks down in front of a rear section of the receiver. This form of locking is used today only in rifles of small caliber and low power because of the great distance between the head of the cartridge and the locking point.

As the bolt is pulled back, the extractor in the face of the bolt draws the empty cartridge case out of the firing chamber with it and strikes it violently against the claw of the ejector causing it to pivot and be hurled out the stop of the breech. This front claw of the ejector projects only when the bolt is drawn all the way back.

Minor modifications of this rifle were made and a short carbine model with wooden forestock extending to the muzzle was also issued.

Development

The German government thought so highly of this new development that they arranged banking assistance and turned the great Oberndorf plant over to the Mauser brothers for manufacture of their rifle and for further development.

Shortly thereafter the Mausers were given additional financial help to produce the rifle in quantity for foreign sale. Thus started the concerted efforts of the German High Command to distribute German arms, and with them German methods, ideology and military thought throughout the world.

Except for comparatively short periods during World Wars I and II, Germany as a matter of military policy from that day on used every means and every effort to infiltrate and eventually direct the equipment and military policies of small Nations throughout the world. The Mauser arms, together with those of Krupp, were the prime products for the world spread of German military thought. By producing only the finest arms and by providing instructors, technicians and tacticians the Germans were able to develop military commissions which blanketed Europe, Asia and South America, where even today German arms and ideologies are paramount.

Brother Wilhelm, not being very strong physically, undertook the position of traveling salesman for the organization while Paul supervised the plant. In 1876 they sold 26,000 of these model 71's to China. In 1881 they sold 120,000 to Serbia as indicated in the records of the Mauser company published in the official plant history issued in 1936.

5. MODEL 1873, GERMAN ARMY RIFLE

The 1873 Model is really a converted French Chassepot. At the close of the Franco-Prussian War of 1870 the Prussians seized large quantities of Chassepots.

The breech mechanism was altered to handle the *cartridge* designed for the 1871 Model Mauser; but the arm is not a Mauser production and should not be confused with the arms of that company. It is listed here merely because many collectors encounter specimens which do not tally with the description given herein of the standard German Model 1871; and this notice should serve to provide correct identification.

The 1873 was issued primarily as a cavalry musketoon having an overall length of about 39-inches and a weight of about 8.38 pounds.

An adequate description of the Mauser conversion for the Chassepot has already been given in the patent papers covering the Mauser-Norris 67-69.

6. THE RISE OF THE REPEATING RIFLE

While the Mauser brothers were starting their work, the Spencer repeating rifle with a tube magazine in the stock had been developed in the United States by 1862 and had done terrible execution in our Civil War in 1865. It was followed shortly by the Henry, which soon was purchased by Winchester and altered to their famous Repeater; this rifle carried the cartridges in a long tube below the barrel.

Thirty thousand Winchester rifles were used by Turkey in their war against Russia in 1877-78 and produced such terrible havoc at the seige of Plevna that all military observers abroad saw the military need for magazine or repeating rifles which would increase infantry fire-power.

Little Switzerland acted in 1867 to adopt a magazine rifle though few were delivered before 1869, and her action was promptly followed on a larger scale by Austria-Hungary.

The Austro-Hungarian Fruhwirth 1870 rifle carrying eight cartridges in the tube magazine in the fore-end, stimulated Mauser in the development of a new repeating rifle.

This Fruhwirth system provided a carrier formed like a scoop and mounted below and in front of the bolt. Cartridges fed into the carrier by the spring in the front end of the tube were lifted successively as the bolt was drawn back; and were driven individually into the firing chamber as the bolt was pushed forward.

Paul Mauser did considerable experimental work on this type of magazine and carrier to develop the first Mauser Repeating Rifle. While Mauser and the German Small Arms Commission testing at Spandau realized the desirability of a smaller caliber arm than the standard 11 mm, they concentrated on a repeater to use the then-standard army cartridge because they did not wish to risk the adoption of a new caliber at a time when war seemed imminent on every horizon.

7. MODEL 71-84 MAUSER, THE FIRST SUCCESSFUL GERMAN REPEATING RIFLE

The repeating rifle developed by Mauser from his original bolt action single shot rifle was formally adopted by the German government in 1884. Officially designated as "Infantry Repeating Rifle M. 71-84," it used the self same 11 mm (.433) cartridge as used in the earlier single-shot model.

The new Mauser was adopted after gruelling field tests following trials by the Army Commission which in 1884 recommended the adoption for the army of this rifle. By the spring of 1886 the entire German Army was equipped with it.

The rifle weighed 10 pounds 2.25 ounces (about 14 ounces heavier with bayonet) and measured 4 feet 3 inches over all (5 feet 9.5 inches with bayonet).

The barrel was 31.5 inches long and the caliber was the standard 11 mm (.433). The rifling was 4-grooves of .0079 inch depth with a twist of one turn in 21.65 inches, direction being to the right.

The sight setting ran from 200 meters (218.7 yards) to 1600 meters (1750 yards).

The velocity and range were the same as in the single shot, about 1425 feet per second velocity at the muzzle and a carrying distance of 3280 yards.

The magazine of this rifle was loaded with eight cartridges inserted through the action and pushed forward. The first one inserted pushed the follower forward to compress the spring within the magazine tube. Each successive cartridge inserted pressed against the base of the one ahead of it to further compress the spring. When the magazine was fully loaded, a ninth cartridge could be placed on the carrier (or "riser"), the bolt started slightly forward, and a tenth one inserted directly in the barrel chamber.

Pressing the bolt forward and turning the bolt handle down locked the breech and let the extractor snap around the rim of the cartridge in the chamber. As the motion of the bolt lever cocked the striker, the arm was now ready to fire.

After firing, turning up the bolt handle unlocked the breech, and pulling back on the bolt handle extracted and ejected the cartridge exactly as in the earlier type. The final motion of the bolt in its rearward travel however depressed the rear end of the riser or carrier

MODEL 71-84 GERMAN INFANTRY RIFLE, RIGHT SIDE VIEW WITH ACTION CLOSED

This is the Model 71 action modified to utilize a tube magazine below the barrel to convert it to a repeating rifle. The lockwork is essentially that of the Model 71 single shot.

Photos and drawings are reproduced from early Mauser originals.

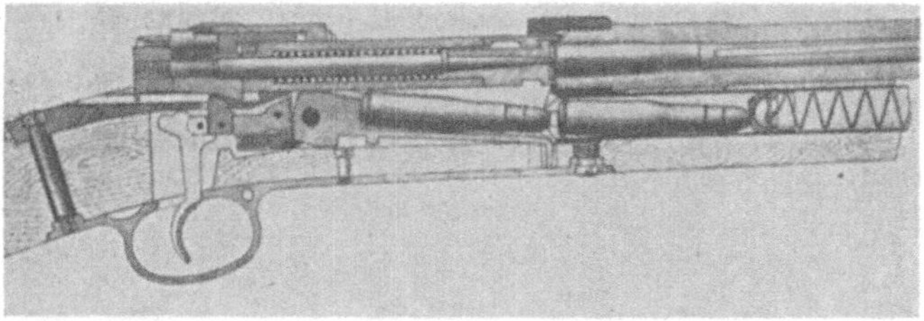

MODEL 71-84. RIGHT SIDE PHANTOM VIEW SHOWING DETAILS OF ACTION AFTER FIRING CARTRIDGE IN CHAMBER

In this position, as the bolt handle is lifted it will first withdraw the striker back into the bolt away from the head of the fired case, this being accomplished by cam action. As the bolt handle is lifted through 90 degrees to the left, and drawn straight back, it will withdraw the empty cartridge case and strike it against the ejector to hurl it out of the rifle.

As the bolt travels back and ejects, it will ride down the head of the pivoted cartridge elevator on which a cartridge is resting. This will lift the front end of the elevator, whose nose will block the cartridge in the magazine from moving forward under the influence of the magazine spring in the tube.

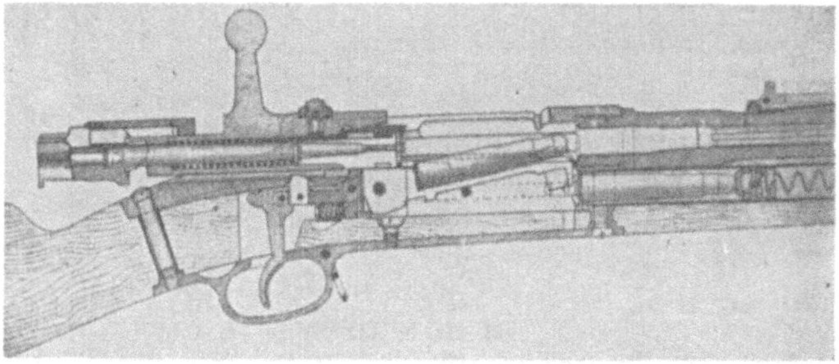

MODEL 71-84. RIGHT SIDE PHANTOM VIEW SHOWING ACTION OPEN AND IN FULL REAR POSITION, EMPTY CASE EJECTED, ELEVATOR RAISED READY TO FEED CARTRIDGE AND TUBE BLOCKED

The bolt head traveling to the rear and forcing down the rear head of the elevator has tilted it to bring the cartridge in the elevator trough up into line with the feeding face of the bolt.

When the bolt is thrust forward, the head of the bolt will start the cartridge forward into the chamber and continued further movement of the bolt will ride over the elevator trough and force it down. When the hooked front end of the trough has been lowered far enough, it will line up with the cartridge in the magazine tube. The compressed spring within the tube, forcing the follower will drive the cartridge into the elevator trough ready for the next lifting movement in feeding.

Closing the bolt will completely chamber the cartridge

and spring the extractor around the head of the cartridge case.

Turning the bolt handle down will act through the standard Mauser cam action to force the cocking-piece back to full cock and enable the sear to engage and hold the striker back ready for firing.

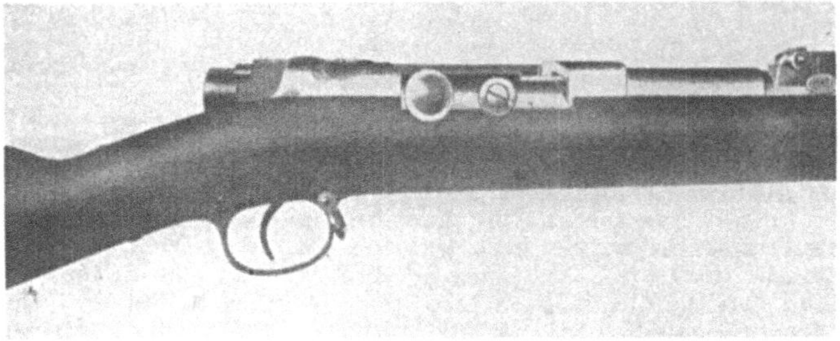

MODEL 71-84. RIGHT SIDE VIEW OF RECEIVER WITH ACTION CLOSED

The design of the bolt way in the top of the receiver has been modified in this design.

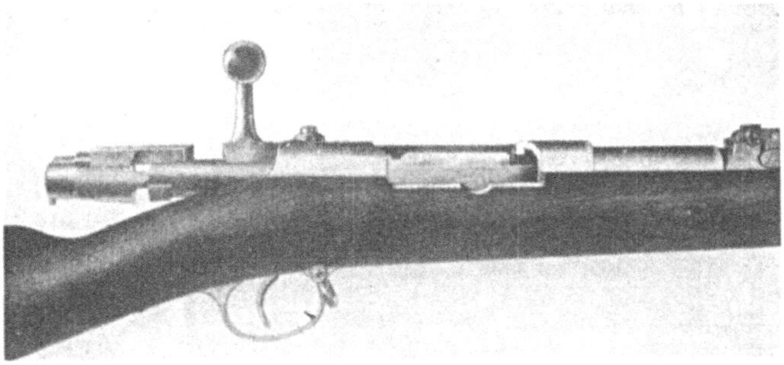

MODEL 71-84. RIGHT SIDE VIEW OF RECEIVER WITH ACTION OPEN

Note that the cocking-piece design and cam surfaces have been modified in this arm.

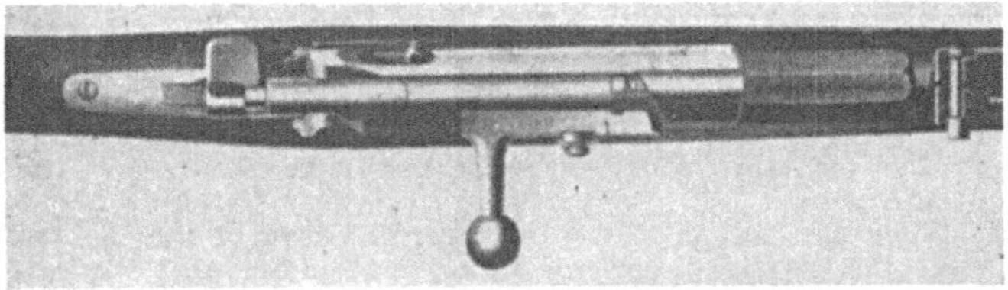

MODEL 71-84. TOP VIEW OF RECEIVER WITH ACTION CLOSED

The left side wall of the receiver was modified in this design from that used in the Serbian model. When the action is open, cartridges can be inserted through the feedway and thrust down into the magazine tube below the barrel.

This design does not employ the side loading gate familiar to those using standard U. S. tubular type magazine rifles.

to line up the cartridge on it with the bolt for loading into the chamber. A hook at the forward end of the riser locked the next cartridge in the magazine tube, preventing it from emerging at that point. Pushing the bolt handle forward stripped the cartridge out of the riser and chambered it. The final closing movement of the bolt depressed the riser and permitted the spring in the magazine tube to force the cartridges back and press the first cartridge in line onto the riser ready for that member's next lifting movement.

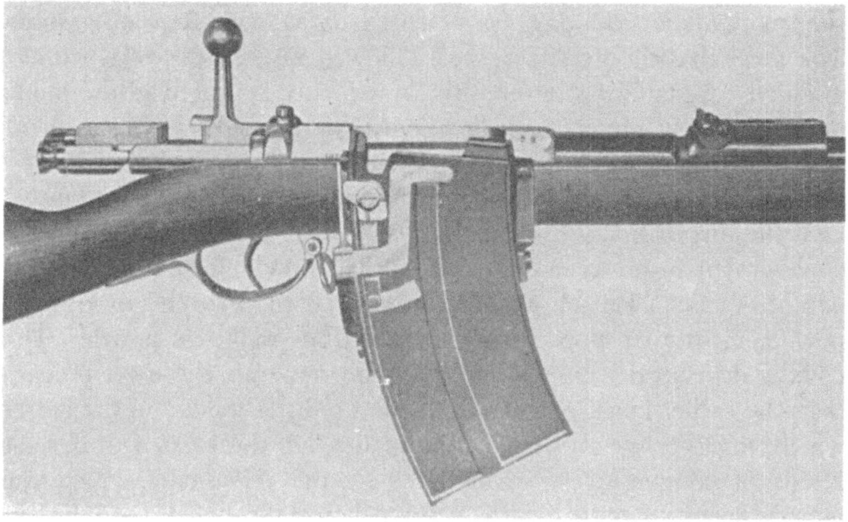

SPECIAL MAUSER BOX-MAGAZINE BUILT ON THE MODEL 71 SYSTEM

This was one of the designs developed by Mauser as an improvement on the tube magazine repeater. Because of the changing balance of the rifle as the magazine was emptied in the tubular type, Mauser conceived the idea of this box magazine (which is curved because the heads of the cartridges had rims) which would be inserted in the rifle from below and could be used to convert the 71 Model to a large capacity magazine rifle. This general type of magazine was later widely adopted and used by practically all nations. The first patent filed on a vertical box magazine was that of the American inventor, James P. Lee, in 1879.

This model was manufactured in tremendous quantities. In 1887, 550,000 rifles and carbines were delivered to Turkey alone. As in the case of the original single-shot Mauser rifle, the 71-84 was made both as a rifle and as a carbine. The carbine had a short upper forestock. This rifle was found being used in quantity in World War II, which provides a commentary not only on the original quality and value of the arm itself, but also on the ability of a country such as Germany to hide, store and preserve arms over long periods of time.

8. DEVELOPMENT OF FRENCH COMPETITION

When the French finally awoke to the fact that the Germans had completely equipped their army with repeating rifles, France undertook the development of a similar magazine rifle. In his "Military Small Arms and Ammunition" (1902), R. H. Angier, one of the outstanding English authorities in the last quarter of the Nineteenth Century writes: "Desultory experiments were made with magazine arms, but no Great Power followed the example of Switzerland until it became known that the German Government had quietly re-armed the troops stationed along its Western frontier with Repeating Rifles. The inevitable counterstroke soon followed in the next year, rumors, gradually crystallizing into positive statements, appeared in the public press, of a new Magazine Rifle adopted by the French Government, and being issued with all speed to the French troops. Everyone will remember the accounts, largely fanciful, given of its wonderful powers, and the precautions, as elaborate as futile, taken by the French Government to preserve the secret of their new ammunition. In 1885 one M. Vielle, a French chemist working at the French Government powder factory developed the first *successful* smokeless powder. This powder developed a higher velocity with comparatively lower pressure than the earlier black powder, and left very little residue in the barrel, which, together with its smokeless quality, led the French to develop a rifle of comparatively small bore, 8 mm (.315 inch) which was more efficient than the earlier heavy bore rifle being used by the German. Indeed this French Lebel, as it was called after Gen. Lebel of French Ordnance, the developer, was the first truly modern small bore rifle."

The German Army Commission experimented to produce a small bore rifle to use smokeless powder to counter this new development of the French. From this developed the next German-adopted weapon manufactured but not invented by the Mauser organization, one which has been the subject of great controversies.

At this juncture it might be well to state that early Mauser rifles, quite like the later ones, were manufactured at many plants besides those of Mauser at Oberndorf. Specimens may be found bearing the marks of the government arsenals such as Amberg, Danzig, Erfurt or Spandau. They will also be found marked "Oesterr. Waff Ges.," many having been made by the Steyr works in Austria. At a very early date liaison between the German Mauser factories and those of Austria and Belgium were already underway. Effective liaison was later to spread to Belgium, Czechoslovakia and Spain; countries which have always filled foreign orders for Mausers when the German factories were either too busy or were closed after wars.

Before passing on to the next German Army Rifle, Model 1888 mention must be made of the Serbian cal. 10.15 and the Turkish 9.5 mm weapons designed and made by Mauser.

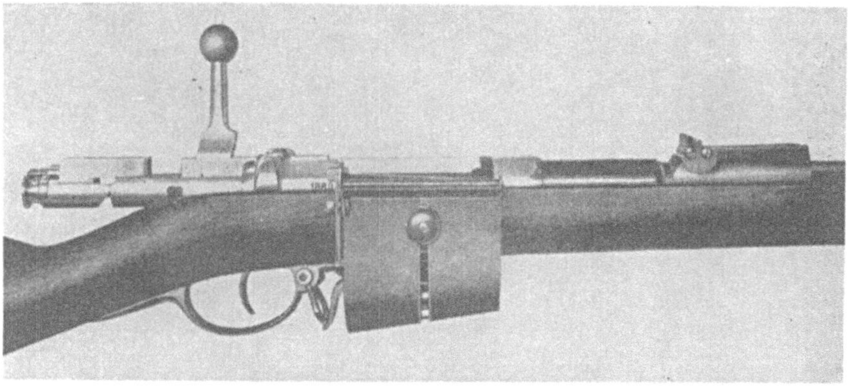

MAUSER EXPERIMENTAL, LARGE CAPACITY MAGAZINE

This was another Mauser design intended to give better balance, easier loading and faster operation than the tube magazine type.

In this design, the box was detachable, being controlled by a spring clip or catch. The magazine itself roughly resembled a horse shoe. Cartridges fed down on top of the follower on the right side followed around the curve and up the left side of the box to give maximum magazine capacity. Note that this unusual design had the compressor button attached to the follower to permit drawing the follower down and compressing the spring to make loading easier. A lip at the top of the magazine prevented the cartridge from being expelled accidentally.

9. THE FIRST SERBIAN REPEATER

Shortly after the German adoption of the 71-84, Paul Mauser designed a similar arm for Serbia in caliber 10.15 mm. 4000 of these were delivered as 7-shot carbines; and 4000 more slightly longer and heavier as 8-shot rifles for use by artillerymen.

Except for the mechanical differences entailed in the caliber and magazine capacity changeovers, these were the same as the German 71-84.

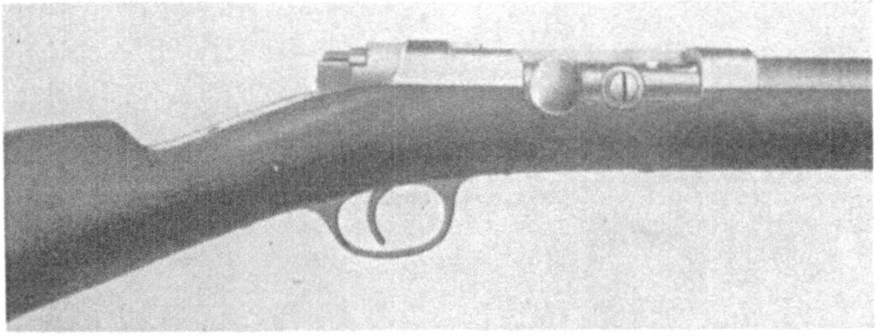

MODEL 78-80 SERBIAN RIGHT SIDE VIEW OF RECEIVER SECTION WITH ACTION CLOSED.

This is merely a slightly modified German 71 type. The receiver bridge and cocking-piece vary in shape and size.

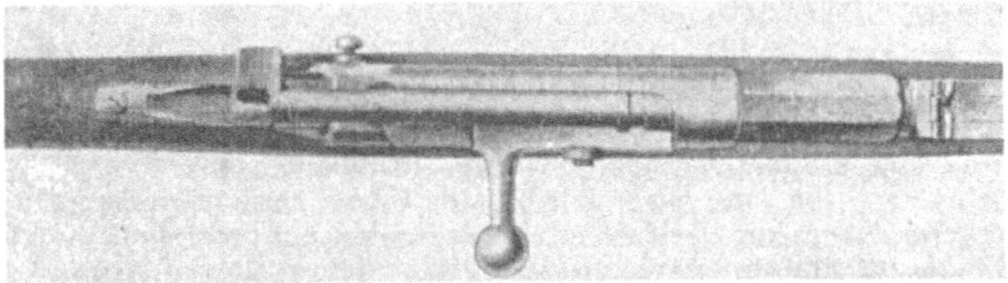

MODEL 78-80 SERBIAN TOP VIEW OF ACTION

The left hand receiver wall in this design more fully encloses the bolt than in the earlier German type. This was one of the important modification of design. It gave added strength to the action and provided for easier loading and better type ejection.

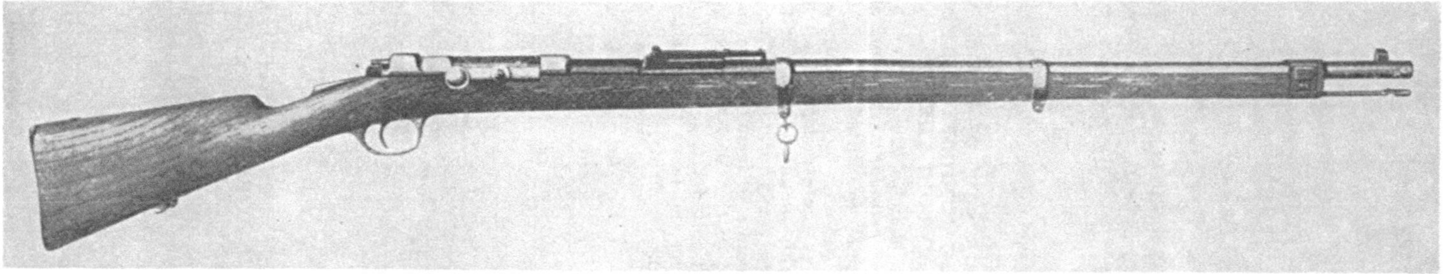

MODEL 78-80 SERBIAN INFANTRY RIFLE.

Right side view of the rifle with action closed.

This design was also made in the shorter carbine form.

This rifle measured 50.7 inches overall and weighed 9.9 lbs.

The caliber was .395 inch, the barrel being rifled with 4-grooves to the right.

The sights were ranged from 300 to 2025 meters.

This cartridge measured 3.08 inches overall and weighed 617 grains.

The length of the case was 2.35 inches and its weight 208 grains. The bullet weighed 340 grains, and measured 1.13 inches overall. Powder charge was 4.8 g of special black powder.

Many of these rifles were encountered in World War II in the Balkans. Some were converted to use box type magazines.

MODEL 78-80 SERBIAN RIGHT SIDE PHANTOM VIEW WITH ACTION OPENED AND CARTRIDGE IN FEEDWAY READY FOR FORWARD BOLT THRUST

Knob of bolt handle is not shown. The striker pin is drawn back into the breech bolt cylinder as a result of the camming action as the bolt handle was lifted.

Pushing the bolt forward will cause its face striking against the base of the cartridge to drive the cartridge forward and chamber it. The extractor movably mounted in the bolt head will snap over the rim of the case ready to extract it on rearward movement. As the bolt handle is turned down to lock into its receiver cut, the turning motion will act through the cam faces on the rear of the bolt cylinder and on the cocking-piece to complete cocking the arm and leave the sear holding the striker in rear position ready for firing.

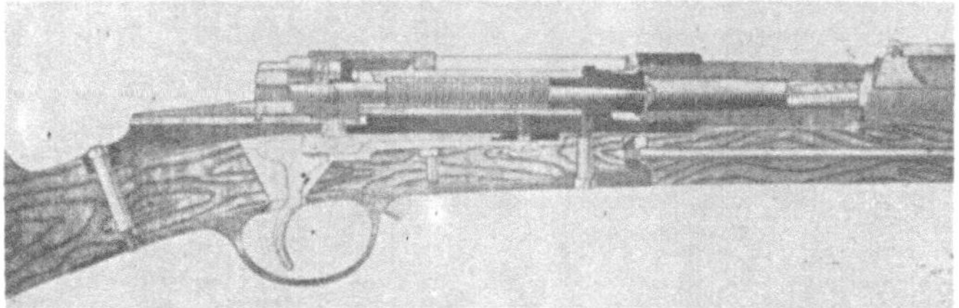

MODEL 78-80 SERBIAN RIGHT SIDE PHANTOM VIEW WITH ACTION CLOSED AND CARTRIDGE IN CHAMBER FIRED

All parts are forward and at rest. Note that this is a single shot rifle. Cam action as the bolt handle is lifted pulls the firing pin away from the primer of the fired case and draws it back inside the bolt.

10. THE 9.5MM TURKISH MAUSER

After intensive experimentation and observation Mauser had found that the best ballistics could be obtained, with the black powder then used, by reducing the caliber to 9.5 mm.

This new rifle with improved locking lugs was offered for Turkish tests. In 1887 Turkey gave an order through Ludwig Loewe & Company of Berlin for 500,000 rifles and 50,000 carbines of this caliber and design—Mauser and Loewe being 50-50 partners in the order.

This rifle was 49.5 inches overall and weighed 9.37 lbs. The barrel was rifled with 4-grooves to the right. Sights were ranged from 200 to 1600 meters. Bore diameter .374. Groove depth .0059.

The cartridge measured 2.47 inches overall and weighed 555.5 gr. The case was 2.34 inches long and weighed 194.46 gr. Bullet length was 1.047 inches, weight 284.9 gr., hardened lead. Diameter .382.

The 1887 Turkish is very seldom encountered, as the cartridge for it was in manufacture for only a short period, and obsoleted rifles were widely destroyed by the authorities.

MODEL 87 TURKISH TUBE-MAGAZINE RIFLE

Right side view with action closed. This is the 9.5 mm caliber of Mauser design. The tube magazine is in the wood stock below the barrel.

Except for changes as required to handle this smaller cartridge, this arm is the same as the German Infantry Rifle, M71-84.

Today this model is a rare collector's piece. Photograph from contemporary Mauser record.

11. MODEL 88, GERMAN INFANTRY RIFLE

The German Rifle Model of 1888 has been the subject of controversy among gun experts for years. Because the rifle was loaded with a clip resembling that of the Mannlicher and based on his patents, it is often mistakenly referred to as the "Mannlicher" or as the "Mauser-Mannlicher." Its official designation was "German Infantry Model 1888"; but it is often listed in Germany as "Mauser and Commission."

This rifle was actually a development of the German Infantry Board or Commission and no receiver of this model ever bears the Mauser factory stamp. These arms were manufactured in huge numbers in a Mauser plant, but in 1888 finances of the Mauser Company had been taken over by the giant German firm of Ludwig Loewe and Company of Berlin and such rifles are usually stamped "Loewe, Berlin 1888, 1890, 1891" or simply "Gew. 88."

Model 88's were also manufactured by all the standard government arsenals, and by such well-known makers as Haenel, Schilling, Sauer and Sohn at Suhl; and by Steyr in Austria. In 1889 arrangements were made for manufacture of rifles of this type in Belgium by Auguste Schriever et Cié at Liege. In the official Mauser plant record "Mauser Gewehr und Patente," published in 1936, is found the statement that Mauser submitted design ideas to the Commission and was very annoyed when they passed over his design recommendations. Indeed, Paul Mauser was extremely bitter about the adoption of this rifle, contending it was inferior to his newer experimental designs. Time, incidentally, proved that Mauser was right.

In the early official catalogues of Ludwig Loewe and Company, manufacturers of genuine Mausers during that period, is found the following statement: "In this rifle (Gew. 1888) there was retained the approved breech closing mechanism of the Mauser rifle pattern 1871 with slight alterations, and there was combined within the cartridge holder from the Austrian Mannlicher rifle.

"This latter feature, however, was not adopted without an important alteration made by the Small Arms Inspection Committee in Spandau, which alteration made it possible to arrange or pack the cartridges symmetrically, and introduce the packet into the magazine with either side uppermost. The cartridge holder constructed by Mannlicher up to that date had a rhomboid form and could therefore only be introduced into the magazine in one particular way *i.e.,* the holder

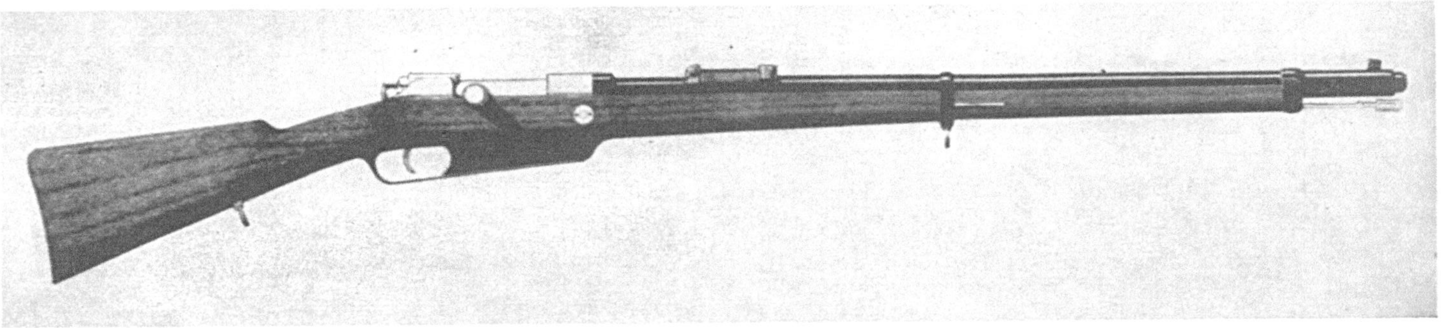

MODEL 88 GERMAN INFANTRY RIFLE. RIGHT SIDE VIEW SHOWING ACTION CLOSED

This rifle is *not* a Mauser. It is included in this book because tremendous quantities were manufactured by Mauser and by Ludwig Loewe and Company who controlled the Mauser finances, at the time of manufacture.

This rifle was developed by a German Army Commission, and Paul Mauser was particularly bitter about its adoption. While the bolt is modified from the original Mauser design, the magazine is a modified Mannlicher, as is much of the rest of the arm.

This rifle used the 7.9 mm cartridge developed by Germany as an outstanding advance in ballistics in its day.

The M. 88 cartridge was used also in Mauser's improved "Gew. 98." In 1905 the case was retained but the bullet diameter was increased. U. S. commercial "8 mm (7.9 mm) Mauser" cartridges may be used.

Modern cartridges of German 7.9 mm military caliber with pointed bullets will chamber in this rifle but are very dangerous to use.

Arms of this type should be loaded only with comparatively low velocity loads expressly manufactured for use in them.

had an upper and a lower side, and the lower side must always enter the magazine first."

Among the other "minor modifications," so called, was the use of front locking lugs on the bolt which provides the most secure breech locking system ever developed.

This system of forward locking lugs had been used in the United States on a breech loading, bolt action, cap-and-ball rifle developed by Colonel J. Durrell Greene of the U. S. Army. This rifle patented November 17, 1857 was unsuccessful because metallic cartridges which would seal the breech against escape of gas had not been perfected at that time.

The Packet-Loading (or Charger-Clip Loading) System

The Mannlicher system of packet loading was introduced to speed up loading, as one motion charged the magazine.

In this Mannlicher system of packet or multiple loading, 5 (or more) cartridges are held together fairly parallel to each other by a clip of sheet metal which covers the rear sides of the cartridges for approximately half their length, and fully encloses and guides the cartridge case heads. The fully loaded clip is placed in the receiver when the bolt has been turned and withdrawn and the *clip and cartridges as a unit* are pressed into the magazine where they are held down by a latch which engages in a projection on the back of the clip.

In operation, the bolt functioned in approximately the same manner as all the earlier Mauser rifles. When the handle was turned up and drawn back the packet could be inserted. Pushing the bolt forward stripped the top cartridge from between the lips of the clip (or "packet") and chambered it. The extractor snapped into the cannelure of the new cartridge, which was rimless.

After firing, the bolt was raised to unlock, then pulled back to extract and eject the empty case. A spring-controlled follower in the form of a lever was forced up by a spring against the bottom cartridge in the clip. The top and bottom of the clip, while cut away and folded over with lips to retain the cartridges, is open enough to permit the magazine follower to ride up between the clip sides.

This lever acted on by the spring forced each cartridge up successively into line as the bolt was drawn back. When the last cartridge had been driven into the chamber, the clip was free to fall out through a hole cut in the bottom of the magazine well. As long as there were

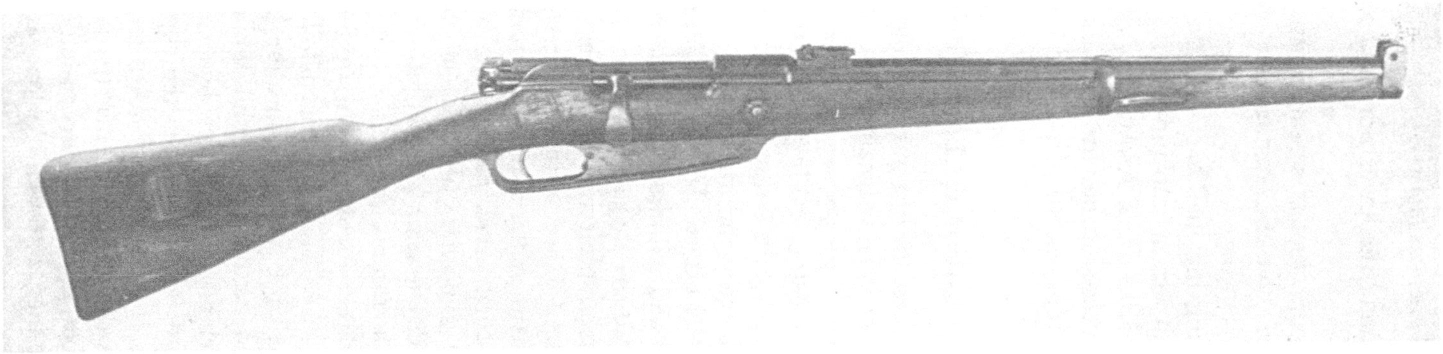

THE 1888 GERMAN SERVICE CARBINE

Note special turned-down bolt handle characteristic of this model. The bolt handle locks down in front of the receiver bridge which is split to allow rearward bolt handle movement. All modern Mausers except the old Portuguese Mauser-Vergueiro have bolts which lock down to the rear of the cylindrical bridge.

Photograph from specimen encountered in World War II. Note that modern German 7.9 mm military cartridges are potentially dangerous to use in this design.

any cartridges in the clip, it was necessary to release the catch by pressing on the thumb piece at the front end of the trigger guard to unload. This released the clip and its contents to be pushed up out of the action by the follower. The hole in the bottom of the magazine was not long enough to permit a loaded cartridge or the clip with cartridges in it to drop out. Only the *empty* clip could pass through it.

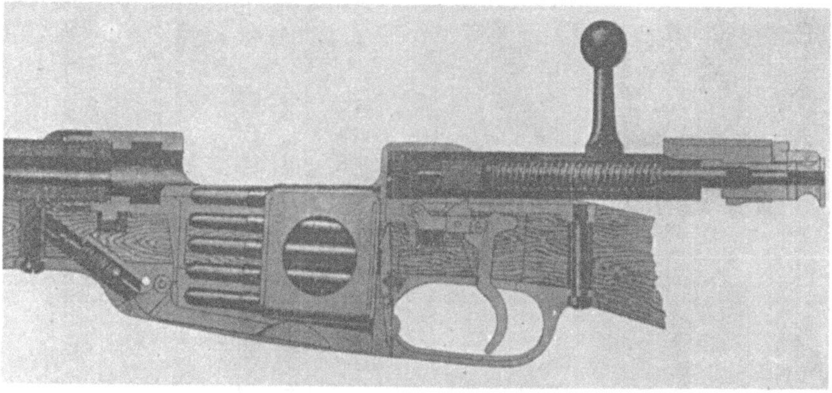

MODEL 88 GERMAN INFANTRY RIFLE. LEFT SIDE PHANTOM VIEW WITH ACTION OPEN AND MAGAZINE LOADED

This rifle, manufactured by Mauser as well as others (but *not* of Mauser design) functions as follows: Lifting the bolt handle through 90 degrees to the left cams the firing pin point back inside the bolt cylinder and revolves the lugs at the forward end of the cylinder out of their recesses in the receiver. Initial extraction is also started by this movement.

The bolt handle is pulled straight to the rear as far as it will go. A Mannlicher type clip of modified design holding 5-round nose Model 88 cartridges is thrust down into the magazine through the top of the action. Since the top and bottom of the clips are open, the bottom cartridge rests on the follower arm and forces it down. As the follower pivots, its forward point forces the magazine spring guide ahead and compresses the spiral magazine spring. When the clip is all the way in, the clip latch is forced forward by its spiral spring to hook into an engagement notch in the back of the clip.

The drawing shows the rifle ready for the forward movement of the bolt. Pushing the bolt forward will cause its face to strip the top cartridge from between the folded-over lips of the clip and chamber it. The extractor in the face of the removable bolt head will snap into the extracting groove in the cartridge case. Turning the bolt handle down will induce cam action to complete closing movement. The cartridge will chamber, and the bolt lugs will be revolved into their recesses behind the head of the cartridge case. The cam surface at the rear of the bolt cylinder acting on the corresponding face in the cocking-piece will force that piece back to complete cocking the arm ready for firing.

This rifle design has several defects. It can normally be used as a single loader only when the magazine is empty. Furthermore the magazine cannot be reloaded by introducing individual cartridges;

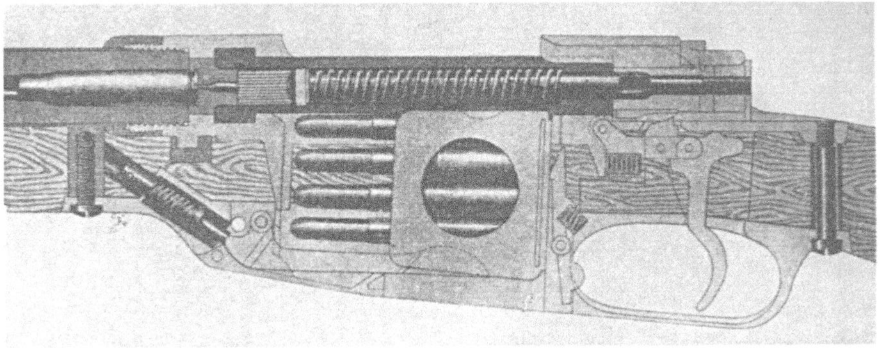

MODEL 88 GERMAN INFANTRY RIFLE. LEFT SIDE PHANTOM VIEW SHOWING ACTION CLOSED AND CARTRIDGE IN FIRING CHAMBER DISCHARGED

The drawing shows the arm after the trigger has been pulled, causing it to pivot and release the sear from engagement with the striker. The striker spring driving the striker forward forces the firing pin point through the hole in the breech-block face to fire the cartridge.

As the cartridge is chambered, the compressed magazine spring acting through its guide forcing against the nose of the magazine follower tilts that member and causes its arm to force the remaining 4-cartridges in the magazine up in line to bring the top one directly below the bolt for the next feeding stroke.

Note that while the locking lugs in this design are at the forward end of the bolt cylinder and are securely locked in their recesses in the receiver, the head of the bolt is a removable piece and hence the lugs are further back from the head of the cartridge than in later genuine Mauser construction.

It is *not* correct to speak of this arm as a Mauser even when it was made by that company. The arm is never so designated in any Mauser records. Popularly it is known as the "Commission and Mauser" because some of the construction details were adapted by the Commission from the Mauser design. Its official German designation was "Deutschland Infanteriegewehr M. 88" (German Infantry Rifle Model 1888).

it ***must*** be clip loaded, since the clip itself is an actual part of the magazine.

A partially filled Mannlicher clip in the action makes it impossible to load the chamber or magazine with a spare cartridge. In the Mauser system, the chamber can be loaded by pushing the cartridges down in the magazine, easing the bolt forward over their heads, then inserting a cartridge in the chamber and closing the bolt. A partially empty Mauser magazine can be filled with single cartridges anytime the action is open. The essential difference is that in the Mannlicher the clip *goes in with* the cartridges; while in the Mauser the cartridges are stripped in *off the clip.*

A thin steel jacket (or "barrel casing") around the barrel was intended to protect the rather thin-walled barrel from injuries and to protect the hands of the soldier from being burned by the heat generated during rapid fire.

Specifications

The Model 1888 rifle introduced an entirely new and outstanding cartridge known as the 7.9 mm. (It is also listed as 7.91 and 7.92 mm). That design is the basis of the cartridge used by Germany from then on until the close of World War II. Except for ballistic changes in the shape of the bullet and the type of charge it is the same cartridge—one of the most efficient known.

While the *nominal* caliber is 7.9 mm, both manufacturers and War Ministry publications list a *maximum* diameter for the *bullet* of 8.1 mm for the M. 1888. The bore diameter is listed as 7.9 mm.

This cartridge (popularly known as "8 mm" Mauser in the U. S.) is the German 8x57mm rimless. It was originally issued with round nosed bullet. Bullet weights and styles, as well as loads, have varied with time and place of manufacture. Bullet diameter is about .318 inch.

The rifle measured 48.8 inches over all and weighed 8.4 pounds. The barrel was 29.1 inches long and had a barrel groove diameter of about .320 inch, 4-groove rifling to the right, one turn in 9.45 inches.

Sights were graduated from 500 to 4000 meters.

12. MODEL 88, CARBINES

Three forms of carbines were made in quantity in the Model 1888. One was the rifle itself cut down 10.5 inches and with the bolt handle bent down somewhat. This was really a short rifle.

The second was slightly shorter and had a stock which came to the muzzle. The front band served also as sight guard. This bolt was straight as in the standard rifle.

Number three was the same as number two except that the bolt handle was turned down and a stacking rod was provided.

The 1888 in all its forms was extensively converted, after being discarded by the German Army, to use as a sporting rifle. Because of the soft metal in the receiver, it was possible for European gunsmiths, as well as some American makers of custom rifles, to manufacture beautiful sporting rifles excellently finished, from the receivers or action. Many were converted to employ the Mauser-type box magazine.

Warning. These rifles when found as sporting arms are often known as the "Mannlicher-Haenel" or "Schilling-Mannlicher," because the names of those manufacturers sometimes appear on them. They are converted military rifles, however, not to be confused with the Mannlicher-Schoenauer high grade sporting rifles or the Schilling or Haenel top quality rifles of later design and manufacture.

It must definitely be remembered, however, that these rifles were intended for the 7.92mm cartridge *in its original form.* Later developed German cartridges are much too powerful to be used in this action.

The German sporting rifle cartridge for the 1888 is known as the "8 x 57 rimless." The *actual* caliber is commonly about 7.9 mm, that being the usual German War Office designation for the old "M. 1888" cartridge. In the U. S. Mauser it is generally called 8mm. *All are the same;* but they must not be confused with the 8mm Mannlicher-Schoenauer cartridge which differs in length, shape and ballistics.

1888 rifles can be dangerous because the sharp pointed (Spitzer) bullet ammunition intended for the later models, will chamber. *Under no circumstances should this military pointed bullet ammunition ever be used in the model* 1888 *rifle or in sporters made from the* 1888 *action.*

The sporting model 1888 cartridge uses a relatively light charge of ***modern*** leaf powder with a relatively soft bullet. Since the present model German military cartridge uses more of the same modern powder and a differently shaped bullet of greater diameter, the pressure developed by the new cartridges is dangerous for use in the earlier rifle.

It is true that the M. 1888 was originally reported to use a cartridge with a chamber pressure of about 47,040 pounds. However, pressure testing methods were not scientifically controlled in those days. Furthermore, the bolt in the M. 1888 is bored out from the front to receive a removable bolt face. As a result, the lugs are further back than in genuine Mauser designs. The old bullet diameter of about .318 inch was increased in 1905 to about .323 inch.

Americans who would not think of using a low number Springfield 1903 with modern ammunition, will unhesitatingly so use this old German type which was built while our Army was still using black powder single shot rifles!

The compression of the bolt lugs, their seats and the removable bolt head, require headspace attention they seldom get. These factors can be very serious with these early model German arms when used with modern ammunition, even where the bore has worn down.

In recognition of these factors, manufacturers in recent years have seldom loaded the M. 1888 type cartridge to give a pressure much higher than 35,000 pounds, using a relatively soft bullet.

Much confusing information has been spread through the years about the German round nose and pointed bullet 7.9 ammunition. The diameters of the bullets, length of bullets, chamber pressure and other details as generally published in English are at wide variance with the official Mauser figures.

In general the Chamber Pressure is listed at 21-tons per sq. inch for the *round nose* and 17.5 tons for the *pointed nose*. In America these figures are usually quoted in foot-pounds on a basis of 2000 pounds to the ton, without making allowance for the fact that the British ton is 2240 pounds.

Furthermore, the *official Mauser figures* are 3200 Atmospheres for the *round nose* (an Atmosphere is approximately 14-2/3 foot pounds) and 3500 Atmospheres for the pointed bullet! In other words, 46,933 ft. lbs. for the *round nose* and 51,333 ft. lbs. for the pointed nose.

Since the official Mauser figures allow a greater maximum bullet diameter, *plus* an increased chamber pressure for the modern pointed

(S) bullet, the danger of using such a cartridge in the obsolete M. 1888 rifle should be apparent.

For purposes of comparison, the official Mauser specifications on the two cartridges and on the Gew. 98 as modified for them are reproduced herewith. It must be emphasized that these are *pre-War* figures, and that arms and ammunition made during the War may vary.

Gewehr:	
Gesamtlänge	1250 mm
Gesamtgewicht	4,1 kg
Lauf:	
Kaliber	7,9 mm
Länge des Laufes	740 mm
Zahl der Züge	4
Tiefe eines Zuges	0,15 mm
Breite eines Zuges	4,4 mm
Drall, ein Umgang auf	240 mm
Drallrichtung	rechts
Drallwinkel	5° 54′ 15″
Länge in Laufweiten	30,4
Durchmesser unter dem Korn	15,5 mm
Durchmesser unter dem Gewindeteil	28 mm
Handschutz	Kurzes Schutzholz
Visier:	
Art des Visiers	Richtbogen
Zahl der Kimmen	1
Höhe des Korns über der Seelenachse	21,2 mm
Höhe des Standvisiers	23,3 mm
Länge der Standvisierlinie	645,9 mm
Standvisier auf Distanz von	400 m
Höchste Visierstellung	2000 m

Munition:	
Gewicht des leeren Ladestreifens	7 g
Zahl der Patronen im Ladestreifen	5
Gesamtlänge der Patrone	80,6 mm
Gesamtgewicht der Patrone	23,9 g
Länge der Hülse	57 mm
Gewicht der Hülse	10,9 g
Geschoßart	Mantel mit spitzer Form
Geschoßlänge	28 mm
Länge in Kalibern	3,45
Gewicht des Geschosses	10 g
Querschnittsbelastung	20,8 g
Größter Durchmesser	8,22 mm
Material des Mantels	Flußeisen, kupfernickelplattiert
Material des Kerns	2,5% Antimonblei
Art des Pulvers	Rottweiler Bl. P. 1293
Pulverladung	3,2 g
Ladungsverhältnisse	0,32
Geschoßgeschwindigkeit v_0	875 m
Geschoßgeschwindigkeit v. 25 m	855
Mündungsenergie des Geschosses	390 mkg
Gasdruck	3500 Atm.

Feuergeschwindigkeit:	Gezielte minutl. Schüsse	25
Mechanische Leistung:	Ungezielte minutl. Schüsse	50
Zahl der Gewehrbestandteile		63

Official Mauser dimensions and ballistics of 7.9 mm pointed bullet cartridge. Compare with similar table for the old round nose bullet and Mauser rifle for it as shown on page 87.

Note also that the Austrian 8mm Mannlicher-Schoenauer is an entirely different cartridge from either one of the German. It is actually 8.2 x 56mm. Its case is shorter than the Mauser case, and the neck is differently placed.

Some idea of the number of these rifles made may be gained from the fact that in the year 1890 to 1891, Ludwig Loewe & Company supplied 425,000 model 1888 rifles to the German Army alone. These rifles were manufactured at the two Loewe plants at Charlottenberg on the outskirts of Berlin, and at the Mauser Waffenfabrik, at Oberndorf, in Württemberg. All these plants were under the financial con-

Gewehr:	
Gesamtlänge	1250 mm
Gesamtgewicht	4,1 kg
Lauf:	
Kaliber	7,9 mm
Länge des Laufes	740 mm
Zahl der Züge	4
Tiefe eines Zuges	0,125 mm
Breite eines Zuges	4,4 mm
Drall, ein Umgang auf	240 mm
Drallrichtung	rechts
Drallwinkel	5° 54′ 15″
Länge in Laufweiten	30,4
Durchmesser unter dem Korn	15,5 mm
Durchmesser unter dem Gewindeteil	28 mm
Handschutz	Kurzes Schutzholz
Visier:	
Art des Visiers	Richtbogen
Zahl der Kimmen	1
Höhe des Korns über der Seelenachse	21,5 mm
Höhe des Standvisiers	23,0 mm
Länge der Standvisierlinie	645,75 mm
Standvisier auf Distanz von	200 m
Höchste Visierstellung	2000 m

Munition:	
Gewicht des leeren Ladestreifens	7 g
Zahl der Patronen im Ladestreifen	5
Gesamtlänge der Patrone	82,5 mm
Gesamtgewicht der Patrone	28 g
Länge der Hülse	57 mm
Gewicht der Hülse	10,9 g
Geschoßart	Mantel mit ogivaler Form
Geschoßlänge	31,3 mm
Länge in Kalibern	3,96
Gewicht des Geschosses	14,7 g
Querschnittsbelastung	30,0 g
Größter Durchmesser	8,1 mm
Material des Mantels	Flußeisen, kupfernickelplattiert
Material des Kerns	5% Antimonblei
Art des Pulvers	Rottweiler Bl. P. 0904
Pulverladung	2,63 g
Ladungsverhältnisse	0,18
Geschoßgeschwindigkeit v_0	640 m
Geschoßgeschwindigkeit v. 25 m	620 m
Mündungsenergie des Geschosses	307 mkg
Gasdruck	3200 Atm.

Feuergeschwindigkeit:	Gezielte minutl. Schüsse	25
Mechanische Leistung:	Ungezielte minutl. Schüsse	50
Zahl der Gewehrbestandteile		62

trol of Loewe. The combined plants' capacity was 60,000 rifles per month.

Some idea of the tremendous extent of these manufacturing operations may be gained from the following figures: The rifle barrel required 54 operations to complete. The bolt assembly required 174 operations. The receiver required 120 operations. The complete magazine required 141 operations; while the stock required 49.

13. THE RISE OF THE MAUSER CLIP-LOADER

Shortly after the introduction of the Model 1888 rifle into Germany, Belgium considered adopting a similar type of multiple (or packet) loading arm. However, after careful consideration and tests of loading systems, it was decided that a more desirable form than the Mannlicher would be one in which the cartridges could be inserted singly to reload the magazine when time or opportunity permitted; in which the clip would not constitute a necessary part of the magazine mechanism, since rusting or deformation of the clip could result in jams; and in which the desirable factor of single-motion magazine loading could be retained.

The Belgian Army therefore requested manufacturers of small arms in all countries to submit rifles into which the cartridges could be speedily inserted in groups as in the Mannlicher, but without the drawbacks encountered with the packet system of loading where the arm could not be singly reloaded or the magazine held in reserve during battlefield use. It was stipulated that it was indispensable to the complete reliability of the arm that the magazine should be such as to permit refilling with single cartridges at any time. This principle was upheld in battle experience until the introduction of the U. S. M.1 Rifle in 1936, by which time we had so perfected our supply system that a packet loader was feasible.

From this request of the Belgian Government developed the next Mauser rifle, one based on an invention of Mauser's patented under specification No. 12689 in England in the year 1888. This device was the now familiar cartridge clip upon which five cartridges are mounted so they can be stripped down into the magazine without inserting the clip itself. This system permits reloading with individual cartridges whenever the magazine is partly depleted and opportunity permits. This strip-in clip, by speeding up magazine loading without danger of jamming, completely revolutionized the development of military rifles.

Ironically enough, the first official tests of the new Mauser loading system were held in England; and the British War Office passed up an opportunity to be the first in the field with this new form of speed loading.

14. MODEL 1889, BELGIAN MAUSER

In response to the request of the Belgian Government for the submission of test rifles, weapons were supplied by manufacturers from many countries. After extensive and detailed testing, the Belgian Government finally accepted the new Mauser rifle. This was the first weapon manufactured by Mauser to employ a box magazine, and the first on which his now famous cartridge clip was used as a quick loading instrument.

With only minor modifications, this rifle was officially in use by the Belgian Government in World War I. The essential design of the rifle was so correct that only minor modifications were necessary; and even the most modern Mausers vary from it in detail rather than principle.

Specifications

The caliber of this model is 7.65mm or .301 inch. The overall length is about 50 inches. The barrel measures 30.7 inches. The rifle weighs about 8.5 pounds without bayonet, the bayonet adding an additional pound.

The arm may be loaded with individual cartridges or may be loaded with 5-rounds directly from the clip. No cut off is provided, but the magazine may be charged at any time by opening the action and inserting cartridges. The rifling is 4-grooves to the right with a twist of 1 in 9.842 iches. The sights are rear notch and front barleycorn. The range of sight graduation is from 500 to 2000 meters. No wind gauge is provided. This rifle is noteworthy also for the introduction of the one-piece bolt with full-forward locking lugs, the strongest system known. The overall length of the rimless cartridge according to Loewe is 3.07 inch, while the length of the rounded-nose cupro-nickel bullet is 1.2 inch. The loaded cartridge weighs about 441 grains. The bullet with a nickel jacket weighs about 229 grains. The original powder charge was 47.4 grains of smokeless but this charge varies in later loadings as powder efficiencies increase. The average standard today is about 42.5 grains to give the same ballistics as the original load.

The muzzle velocity is about 2034 feet per second with a chamber pressure of about 39,400 pounds per square inch. Bullet diameter is .31 inch maximum. (Note: Current loadings for this cartridge, as well as bullet styles, vary with time and place of manufacture).

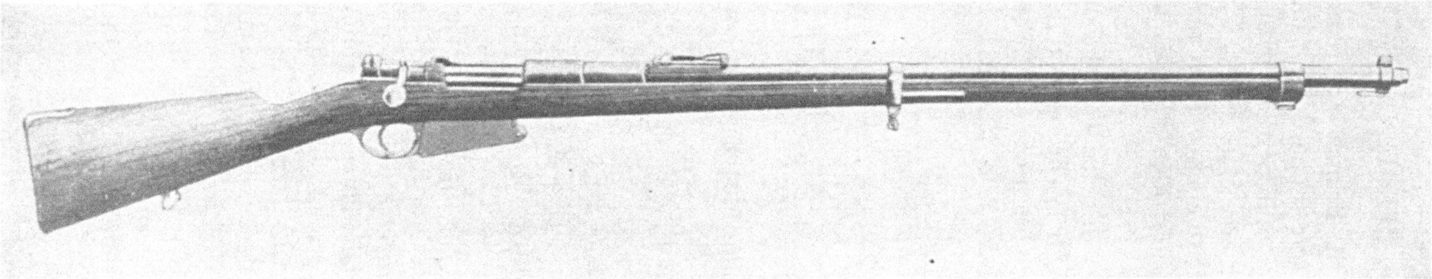

MODEL 89 BELGIAN INFANTRY RIFLE. RIGHT SIDE VIEW WITH ACTION CLOSED

This is the standard rifle officially adopted by Belgium after experimenting with the Model 88 German already shown.

Note that in this weapon a barrel jacket is employed which covers the barrel almost to the muzzle. It was designed to protect the hands of the user from heat during firing and to protect the rather thin barrel from injury in field service.

The magazine is considerably modified from the earlier design, and is adapted to rimless cartridges.

This is still the basic Belgian rifle, though it is now being replaced. Fabique Nationale in Belgium made this model in quantity.

Photograph from original Mauser design records.

Barrel Characteristics

The barrel of the Belgian Mauser differs from that of most of the others in that externally it is left rough from the turning tool. The barrel is quite thin and its diameter at the breech end is about one-inch greater in front of the chamber and again about the center of the chamber. It is screwed into the receiver in standard fashion.

The particular characteristic of the Belgian Mauser is the thin tube of solid drawn steel which covers the barrel and is intended to protect it and also to protect the hands of the riflemen by diffusing the heat

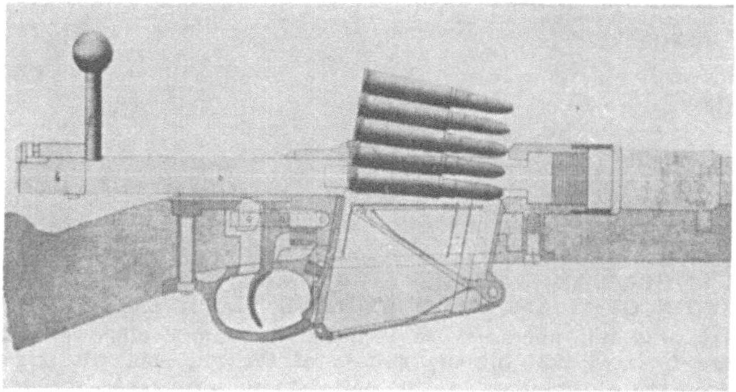

MODEL 89 BELGIAN MAUSER. RIGHT SIDE PHANTOM VIEW SHOWING BOLT IN FULL REAR POSITION AND LOADED CLIP OF CARTRIDGES INSERTED IN CLIP GUIDE READY FOR STRIPPING INTO THE MAGAZINE

The details of the barrel jacket construction may be seen. The type of magazine follower and spring was altered considerably from the original concept in the interests of simplicity of design.

Pressing down with the thumb on the top cartridge will force all the cartridges down out of the grip of the sides of the clip, and by bearing down on the follower and its lever will compress the flat spring below it to provide motive power to force the cartridge up during rearward operation of the bolt. The magazine in this type protrudes below the bottom of the receiver well.

when using the weapon in rapid fire. A ring shrunk about the muzzle end of this tube carries the block for the front sight and the cleaning rod holder. A collar at the front end of this ring provides a mounting for the cross-bar of the sword type bayonet.

A somewhat thicker cylinder is fastened by the screws at one end to the rear of this casing and at the other to the receiver. A pin extends downward from this into a bushed hole in the stock. The rear sight is brazed onto the casing just ahead of the front reinforce. The sight resembles that of the Lee-Enfield rifle except that the V-notches in the slide and the cap are smaller.

A groove is cut on either side of the boltway to permit the working of the bolt lugs and the extractor. The right side groove is partly cut away to permit ejection of cartridges to the right. The recesses for the locking lugs on the bolt are at the front end of these grooves.

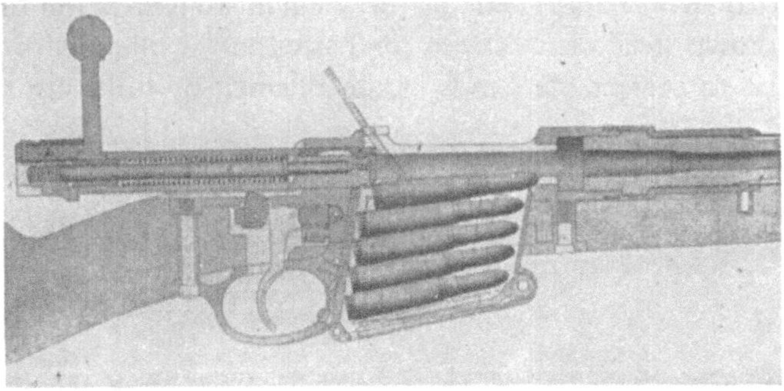

MODEL 89 BELGIAN MAUSER. RIGHT SIDE PHANTOM VIEW WITH ACTION OPEN AND BOLT STARTED SLIGHTLY FORWARD

Note that it is not necessary to remove the empty clips manually. The bolt moving forward will hit the bottom of the clip and tilt it out of the weapon. Continued pressure on the bolt handle will cause the face of the bolt to strip the top cartridge forward and chamber it. Turning the bolt handle down when the forward action is completed will lock the lugs at the forward ends of the bolt cylinder securely in their receiver recesses and will complete by cam action the cocking of the rifle as already described.

As the bolt rides over the magazine, the magazine spring acting through the follower can force the cartridges up in line ready for the next loading stroke.

The rear part of the receiver forms a complete cylinder in which the bolt moves; while beyond this cylindrical point it is prolonged into a tang provided with a groove for the cocking stud.

The receiver is further cut out below to receive the magazine, and there is a cut in the tang to receive the tooth of the sear where it projects into the groove for the cocking stud.

Bolt Design

The bolt itself is exceptionally strong and of very simple construction. It is made in one piece, hence the bolt head is not removable as in early models and in most Mannlicher designs. No tools are needed to dismount it. The locking lugs at the front end of the bolt are placed opposite one another, providing a breech lock directly and firmly behind the head of the cartridge case. This is the most substantial form of breech locking ever devised. All later important military rifles with the sole exception of the British utilize this form of locking. The British "Pattern 1914" used Mauser-type locking, but all the earlier Lee-Enfields and the new Marks of British service rifles retain the old Lee rear lug system.

The face of the bolt is recessed to receive the base of the rimless cartridge. The bolt handle projects at right angles from the rear of the right side of the bolt and ends in a round knob.

A cam-shaped recess at the back end of the bolt serves to partly cock the striker when the bolt handle is turned up; a thrusting movement being imparted to the cocking-piece and striker which cannot turn, as they are acted upon by the turning bolt cylinder. On the opposite side there is a small notch for the tooth of the safety bolt.

The mainspring if of coiled wire, usually 28 coils, measuring .06 inch in thickness.

The striker is provided with a short point and a collar against which the mainspring bears for cocking. The end of the striker is threaded to receive the cocking-piece.

On the under side of this cocking-piece is a projection traveling in a groove cut for it in the tang of the body. This projection engages with the sear nose when the rifle is at full cock. Its front top surface is beveled off to work in a cam slot in the rear of the bolt.

The bolt plug screws loosely into the rear end of the bolt with the mainspring bearing against the front end. The cocking stud working in a slot prevents the bolt plug (or cocking-piece) from revolving with the bolt when the bolt lever is raised.

A rib on the striker prevents the striker from turning in the bolt plug. There is a cylindrical hole in the top of the bolt plug to receive the stem of the safety bolt. The front of the stud on the cocking-piece rests in a small groove at the rear of the bolt and prevents the bolt plug from turning should the safety bolt not be holding back the cocking-piece.

The safety bolt itself is a thumb-piece and a spindle working in the hole in the bolt plug. When the rifle action is closed, and the thumb-piece turned vertical, the flange on the safety bolt is pushed up in front of the top of the cocking-piece to force it back a short distance and enable it to withdraw the stud from contact with the sear. When the thumb-piece is turned to the right, the cocking-piece is still locked,

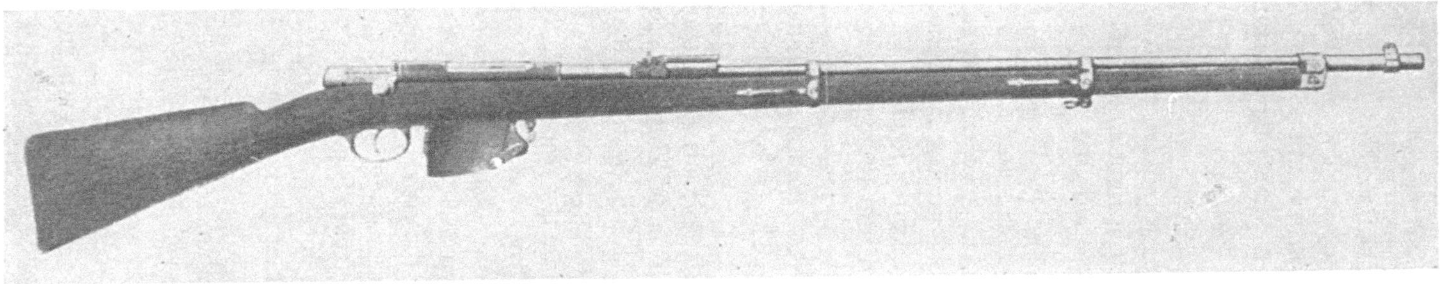

MODEL 88-89. THE ORIGINAL BELGIAN MAUSER. RIGHT SIDE VIEW WITH ACTION CLOSED

This is the first service Mauser rifle issued with a box type magazine. The caliber is 7.65 mm. This rifle marked a turning point in the development of the famous line of Mauser magazine rifles. It is the first to employ the solid one-piece bolt whose head is not removable. The locking lugs at the front end of the bolt are placed opposite one another to give maximum support to the head of the cartridge case.

With the exception of the British service Enfield rifle, and the Swiss, all later military rifles utilize this general form of breech locking.

This model is a collector's item. Photograph from original Mauser records.

but the end of the stem where it is not cut away enters into the slot at the end of the bolt, locking the bolt and plug together to prevent the bolt from being revolved.

This safety bolt is retained in position by a small pin actuated by a spiral spring which works in a groove cut across the top of the bolt plug. The spring and pin are retained in position by a screw.

Extraction System

The extractor is a short spring terminating in a claw. It is mounted into the right side of the bolt in position between the two locking lugs and as the bolt revolves the extractor turns with it. This differs from the other designs of the same manufacturer.

MODEL 88-89 BELGIAN MAUSER. RIGHT SIDE VIEW OF RECEIVER SECTION WITH BOLT FULLY RETRACTED

This weapon was designed for quick loading through the top of the action with the now famous Mauser strip-in clip. The clip was inserted in the guide, and the cartridges were stripped down off it by thumb pressure directly into the magazine below. Note that the magazine was also removable to permit insertion of larger magazines and for cleaning.

The bolt stop is in the form of a hinged lever with a tooth projecting into a groove in the receiver for the left locking lug of the bolt. This lever is pivoted and is kept pressed against the receiver normally by a flat spring which is led into it. When it is pulled away from the receiver by drawing on its flap, it pivots and permits the bolt to be withdrawn from the rifle. (In this model a flap formed to slide against the side of the receiver replaces the rib found in most Mausers).

The ejector is a flat piece inside the retaining bolt. It works on the same pivot pin and is actuated by the spring inside the lever. It projects into the boltway. A slot is cut for it in the left locking lug and in the face of the bolt itself. Thus when the bolt is drawn back, the ejector springs into this slot so that the base of the cartridge strikes against it. As the extractor continues to pull back the right side of the cartridge, this combined motion swings the cartridge case around and throws it out of the action to the right.

The sear projects into the groove cut into the tang of the receiver and is part of a bar pivoted at its front end to the receiver. It is operated by a spiral spring led into it whose other end bears against the receiver. The spring in this type is horizontally placed.

The trigger is pivoted to the bar portion of the sear in which there is a vertical slot. No half cock is provided. The striker cannot be placed at full cock except by opening and closing the action. The pull off is double in standard military practice. The first pull is a sliding movement. This permits the slack to be taken up ready for the actual releasing pull. When this first movement is completed, the leverage shifts. On continued pressure the trigger and sear move to free the seai from the striker bent. The striker is then driven forward by its compressed spring.

Magazine Construction

The magazine holds 5-rimless cartridges *in a single column.* It is a short steel box whose bottom is hinged to the sides at the rear end fastened by a screw at the front end. This screw acts as a pivot for a flat lever actuated by a flat spring attached to the bottom of the magazine. A second flat lever is hinged to the end of the first one and forms the follower for the cartridge (the platform on which the cartridge rests). A second flat spring attached to the lower lever raises the upper one.

The magazine sides are turned in at the top almost their full length to hold the cartridges in the box. Horizontal and vertical cuts in them permit the sides to act as springs of sufficient elasticity to spread when the cartridges are forced down in the magazine, yet are sufficiently strong to held the cartridges from being pushed out by the rising follower. The magazine passes up through an opening in the prolongation of the trigger guard where it is held in position by a small lever catch pivoted to the guard and operated by a small spiral spring. This magazine projects below the bottom of the stock.

This Belgian pattern is not provided with the customary rib on the magazine platform to serve as a bolt stop when the last cartridge has been fired.

The stock is one piece. The stock, receiver and trigger guard are fastened together by screws. The stock is screwed into a boss extending at right angles to the axis of the barrel which serves to pass on the recoil of the barrel and receiver to the stock, thereby saving the screws from being broken by the shock.

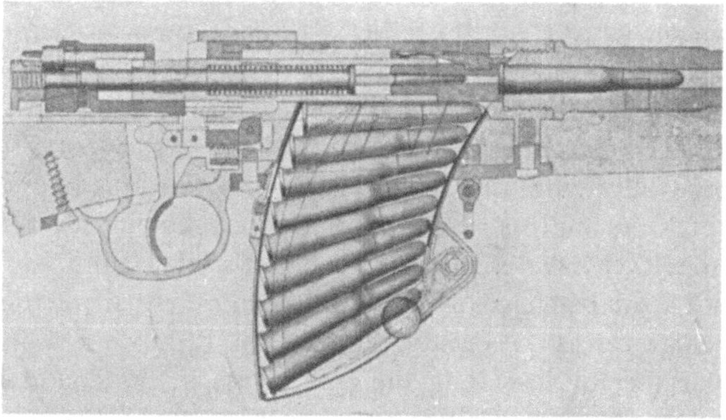

BELGIAN MAUSER WITH 10-SHOT MAGAZINE. PHANTOM RIGHT SIDE VIEW WITH MAGAZINE LOADED AND CARTRIDGE IN CHAMBER READY TO FIRE

The curved construction of the magazine is required by the fact that the 7.65 mm cartridge originally tested has a rimmed case. A special long extractor is required in Mauser actions as indicated, when using cartridges with rimmed cases. The magazine is detachable.

Drawings show the magazine loaded to capacity and the action at full cock ready for the trigger pull.

A projection beneath the upper band is provided to attach the short bayonet. A full length cleaning rod screws into a nut in the stock. Its head has a slot in it which fits into the rod holder of the front sight ring.

The butt plate is steel and is fastened with two screws. Normally it is not provided with a trap.

15. OPERATION OF MODERN MAUSER SYSTEMS

A study of the functioning of the first Belgian model covers in general all details of operation of future models with minor exceptions which will be noted in the text. Mauser's original design was so fundamentally correct that it has never been possible to do more than modify it, and as in our own Springfield (which is a Mauser action), to improve, or more correctly, to refine it to a degree.

As the bolt lever (handle) is lifted to open the action, the cocking stud projecting into the groove in the tang of the receiver prevents the cocking-piece and the bolt plug behind it from turning with the bolt.

The tooth on the cocking stud is thrust back by the cam recess on the bolt thereby drawing the striker back and partly compressing the mainspring. During this raising movement of the bolt lever, the lower end of the lever where it joins the bolt thrusts against an inclined plane cut on the rear face of the cylindrical bridge part of the receiver. This leverage forces the entire bolt assembly to move a short distance to the rear thereby loosening the empty case in the chamber to provide primary extraction (loosening the expanded case in the firing chamber).

After the bolt lever has been lifted its entire travel distance (through an angle of 90 degrees) it ends in a vertical position. The end of the cocking stud now rests in the notch in the rear of the bolt; while the locking lugs have turned out of their locking seats in the receiver and are in the travel grooves in the receiver, permitting the bolt to be drawn to the rear.

As the bolt is pulled back, the ejector springs into its groove. Since the extractor in the face of the bolt is drawing the empty cartridge case back with the bolt, the ejector strikes the opposite lower face of the cartridge case swinging it to the right and thereby freeing it from the grip of the extractor and hurling it out of the action.

The left bolt lug at this point comes in contact with the tooth on the bolt stop, halting the rearward motion of the bolt.

The magazine springs force the next cartridge in line up into the path of the bolt.

Loading and Firing

When the bolt is thrust forward, the ejector is pressed to the left. The bottom of the bolt face strikes the top cartridge in the magazine and pushes it ahead into the firing chamber. As the cartridge is forced ahead, its base is compelled to rise up the bolt face until the extractor catches in the cartridge groove.

When the bolt is about one inch from closing, the cocking stud is engaged by the sear. This results in the cocking-piece and the striker being *held back,* while the bolt and bolt plug *are pushed forward.* This action compresses the mainspring completely. (Note: This cocking system applies *only* to the early Belgian, Spanish and Turkish types. In German and other late types, *turning down the bolt handle* completes cocking.)

This final motion of course completes thrusting the cartridge forward into the chamber.

As the bolt handle is turned down to the right, the bolt cylinder is revolved and the two lugs at the front end are turned into engagement in the recesses cut for them in the receiver directly behind the head of the cartridge. Pulling the trigger levers the sear away from the cocking stud.

The striker is now released and the tooth on the cocking-piece enters the cam recess on the bolt permitting the striker freedom to reach the primer when released.

If the bolt is not completely closed, the travel of the striker is blocked, since the tooth of the cocking stud is not opposite its entering recess. Hence the rifle cannot be fired.

All Mauser rifles wherever they are made or under whatever name manufactured are merely modifications of this original Belgian design.

16. MODEL 1890 TURKISH MAUSER

In 1890 the Turkish government adopted a modified form of the Belgian Mauser. This rifle for the 7.65mm (or .301-inch) Turkish cartridge was minus the jacket tube of the Belgian model.

This model together with minor modifications of the later 1898 pattern issued in 1903 and 1905 are all in official Turkish use, as are Kar. 98's and their modifications.

It must be remembered that despite the fact that many countries used "official" rifles, because of the distribution of weapons during World War II equipment of all types may be found in practically all armies in the world today. Thus tremendous numbers of later models of the Mauser rifle using the *German cartridge* will be found in Turkey. All such rifles, wherever made for Turkey, are slight modifications only of the official German.

However the official 1890 Turkish Mauser rifle has the following characteristics: It weighs 9 pounds 1 ounce without bayonet. The bayonet adds about 1 pound 7 ounces to this weight. The overall length is about 4 feet .5-inch without bayonet. The bayonet adds 1 foot 6 inches. The overall length of the barrel is 29.1 inches, and the barrel is rifled 4-grooves to the right with a twist of one turn in 10 inches.

The sights are standard V and barleycorn with graduations from 250 to 2000 meters and without wind gauge.

The rifle may be loaded with individual cartridges or may be loaded with cartridges stripped in from a clip in standard fashion.

The magazine of the original model followed that of the original Belgian type in that it extended below the bottom of the stock and carried 5-cartridges in a single vertical column, and was detachable.

The original Turkish Model varied from the original Belgian in:

1. Cartridge used. While the caliber is the same, the cartridge was somewhat different. This Turkish cartridge measures 2.96 inches overall, is rimless and weighs 370 grains. The bullet is pointed and has a lead core surrounded by a mild steel jacket. The bullet length is 1.06 inch, the maximum diameter .311 inch; weight 154.3 grains.

As originally issued, the cartridge measured 3.07 inch and the bullet 1.2 inch. The powder charge was 47.4 grains of smokeless. More recent loads average about 46 grains of improved powder. The muzzle velocity is about 2001 feet per second while the chamber pressure is about 39,400 pounds per square inch. (Note: Most modern 7.65mm Mauser military calibers will interchange.)

2. Turkish rifles do *not* have barrel jackets.

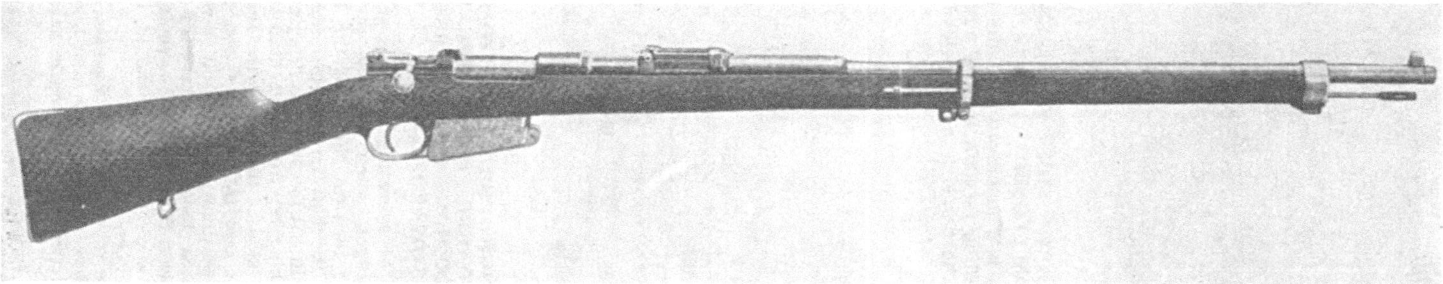

MODEL 90 TURKISH MAUSER. RIGHT SIDE VIEW WITH ACTION CLOSED

This arm is a modification of the Belgian type which eliminates the barrel jacket.

This rifle is still in extensive use among the Turks, but as a first line service arm has been replaced by later improved forms of the Mauser. Tremendous quantities of this model are still in use and in existence however.

While the original Turkish 7.65 mm cartridge differed somewhat from other 7.65 mm Mauser calibers, modern ammunition designated "7.65 mm Mauser" can safely be interchanged.

Photographs from original Mauser records.

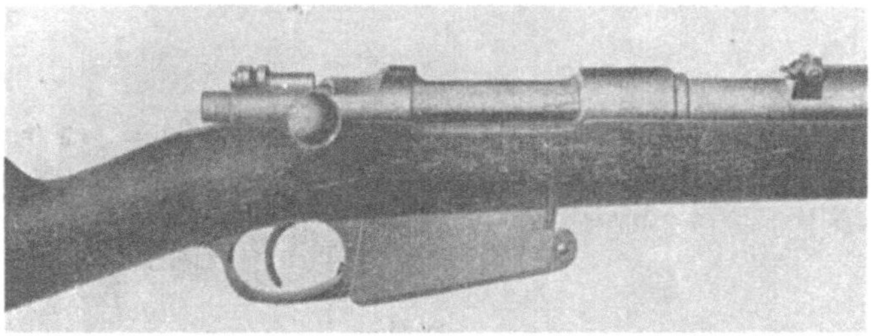

MODEL 90 TURKISH. RIGHT SIDE RECEIVER WITH ACTION CLOSED

Note that magazine design is a modification of the Belgian. This arm is designed for loading through the top with single cartridges or with a clip. The barrel is not provided with a jacket.

MODEL 90 TURKISH. RIGHT SIDE PHANTOM VIEW SHOWING MAGAZINE LOADED, BOLT THRUST FORWARD FAR ENOUGH TO REMOVE CLIP ENTIRELY AND TO START CARTRIDGE TOWARDS FIRING CHAMBER

This demonstrates clearly the operating principle of loading in all rimless type cartridges when worked through a Mauser type magazine. The forward thrust of the bolt has brought the lower edge of the bolt face against the top of the cartridge case and is starting it towards the firing chamber. Note the position of the locking lug directly behind the head of the cartridge case on the side of the bolt cylinder.

As the cartridge is thrust into the chamber and the bolt handle turned down, the rotation of the bolt brings the upper and lower lugs, which have been traveling on the right and left side as they went forward in the receiver well, into locking recesses above and below the head of the cartridge case in the firing chamber.

In magazines using rimless cartridges, the magazine dimensions are smaller and the box can be straighter because of the manner in which the cartridges lie on top of each other.

3. The barrel is turned down in steps with the portion between the steps tapering slightly. The sights are attached to the barrel.

The rear sight is shaped with a thin tube below encircling the barrel and is screwed and soldered to the barrel. The face of the leaf is graduated from 400 to 2000 meters in hundred meter steps. At right angles to the leaf at the rear end is a small projection with a notch for 250 meter fire.

4. There is a second opening in the tang just behind the magazine for a point on the front end of the sear which can rise as the trigger is pressed, to engage in a slot in the bolt.

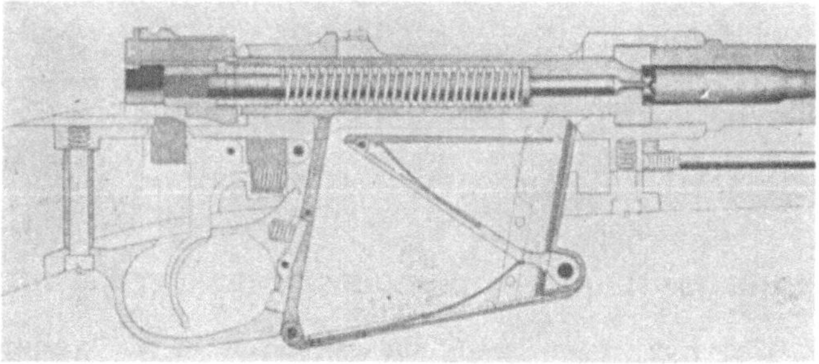

MODEL 90 TURKISH. RIGHT SIDE PHANTOM VIEW WITH MAGAZINE EMPTY AND LAST CARTRIDGE IN FIRING CHAMBER DISCHARGED

The magazine spring acting on the follower lever has thrust the follower arm up to fullest extent. All parts are in complete rest positions.

Firing pin point is still imbedded in the primer of the fired cartridge case.

The first movement of turning the bolt handle up acts through cam action of the cam slot in the rear of the bolt cylinder against the cocking-piece to draw the striker back within the bolt.

Except for a slightly heavier bolt construction, this design is the same as the Argentine Mauser.

5. The cut-off is a small horizontal one pivoted in the right side of the receiver where it is controlled by a flat spring with a tooth which engages in notches in the cut off.

6. A shallow groove is provided in the bolt for the sear.

7. A flap shaped to lie against the side of the receiver is used instead of the conventional rib type bolt stop.

8. The sear differs somewhat, having a projection at the front end engaging in grooves in the bolt and a special spring.

9. The original Turkish rifle has the same type of projecting magazine as the Belgian. However, in later designs (1893 and later)

a simpler and more compact magazine appears; the magazine body forming part of the prolongation of the trigger guard but not projecting below the bottom of the stock.

The opening for the magazine in the receiver in these later models is somewhat narrower than the magazine itself to permit two columns of 2 and 3 cartridges respectively to be pressed upwards against the overhanging edges. This prevents them from rising out of the magazine except in single turn through the central opening.

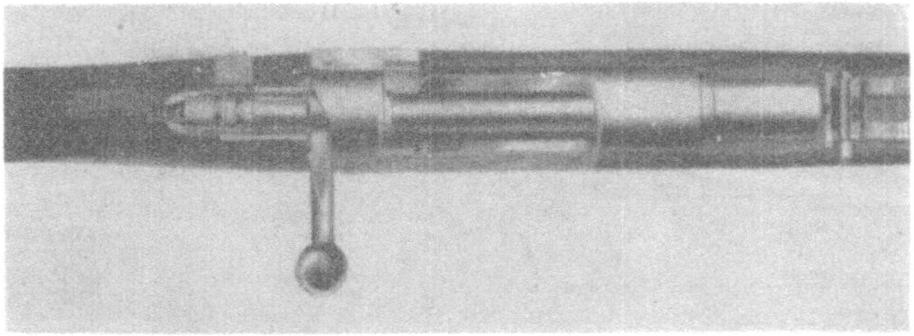

MODEL 90 TURKISH. TOP RECEIVER VIEW WITH ACTION CLOSED

This view shows clearly the cam shape of the receiver bridge against which the side of the bolt handle operates in this design to provide the initial camming action. As the bolt handle is lifted, it must follow the curve in the bridge, and thereby act to withdraw the head of the bolt and provide tremendous primary extraction to start the swollen cartridge case out of the chamber. This extraction is much more powerful than in any other design of rifle. Upward movement of the handle also provides the camming action to force the cocking-piece back far enough to withdraw the striker from the head of the cartridge case. When the bolt is in 90 degree upward position, that is vertical, the lugs are out of their recesses and in their travel tracks where the bolt can be drawn straight to the rear for extraction and ejection.

These principles are the ones still in general use in all bolt action rifles of modern design.

The bottom of this magazine is closed by a plate. The magazine spring is of ribbon steel bent into zig-zag form and has its ends secured in an undercut recess in the plate and in the platform (or follower). A rib on the platform at its left side raises the column of cartridges resting upon it, thereby bringing the center of the cartridges in the left column level with the tops of the corresponding cartridges in the opposite column.

When the last cartridge has been fired, as the bolt is pushed forward its face is caught by the rear end of the rib of the magazine follower, preventing it from being closed until the magazine is recharged or the platform is pushed down by the finger. This serves to notify the firer that the weapon is empty.

10. A cut-off is provided on the Turkish rifle permitting the magazine to be held in reserve as in the case of the United States Springfield while the weapon is used as a single loader.

The mechanics and operations are otherwise substantially the same as those already described for the Belgian Model.

Note on Later Turkish Mausers

These differ but slightly from the corresponding German types, notably the Gewehr and Kar. 98 series. Details covering the German types apply to the Turkish with the exceptions noted.

The official Model 1905 has the following characteristics: Weight is 9 pounds 6 ounces. Length is 4 feet 1 inch (1 foot 8.5 inches more with bayonet attached). Barrel length is 29.1 inches. Caliber 7.65mm Turkish. Right hand twist, one turn in 9.84 inches, 4-grooves. Sights typical V rear adjustable from 400 to 2000 meters and barleycorn front. No wind gauge. Barrel diameter .301 inch. Bullet diameter .311 maximum.

This rifle is equipped with a cut-off permitting holding the magazine in reserve. The magazine is the standard Mauser vertical fixed box, holding 5-cartridges in double staggered rows.

Rifles of Mauser make or pattern to shoot the standard German 7.9 mm cartridge are also common in Turkey.

Germany, Belgium, Yugoslavia, Czechoslovakia and even Poland delivered some Mauser-type rifles to Turkey after 1924. These were in either 7.65mm or German 7.92mm calibers. Mechanically these are the same as the German World War II patterns, since all were based on German minor modifications of the original Mauser Gewehr 98.

17. MODEL 1891, ARGENTINE MAUSER

In 1891 Argentina adopted a slightly modified version of the Turkish rifle. It was improved with a strengthened bolt.

The caliber of this rifle was also 7.65 mm or .301 inch. The length of the rifle was 48.5 inches, and the weight without bayonet about 8.5 pounds. The bayonet adds 1 pound of weight.

The arm could be single loaded or loaded with the Mauser clip of 5-cartridges in standard fashion. The cartridges were retained in the standard magazine in single line.

The cartridge was 3.07 inches long, and weighed about 1 ounce. The bullet was 1.2 inches long, lead with a cupro-nickel jacket and weighed about .5 ounce. The powder charge was about 40.9 grains of smokeless originally giving a muzzle velocity of about 2066 feet per second with a chamber pressure of about 39000 pounds to the square inch. Maximum bullet diameter .311 inch.

The rifling was 4-grooves to the right, one turn in 9.8 inches. Sights were graduated to 2000 meters.

Slight modifications of this rifle and of the ballistics of the cartridge were made, but essentially they remained the same, though in recent years modifications of the German Kar. 98 have also been used.

Later Argentine Patterns

In 1895 Argentina purchased a quantity of rifles from Germany based on the Spanish 1893 Mauser design. Later rifles based on the Gewehr 98 pattern were purchased from both Germany and Spain. Caliber before World War I was uniformly 7.65mm Mauser. Some in 7.92mm caliber have since been delivered.

From 1924 on the Belgian F. N. factory supplied Mauser System rifles to Argentina and other nations. These were all made to the general specifications of the official German Kar. 98 except for such modifications as were required by the difference in caliber. A reading of the chapter on the later German types will effectively cover these later Argentine rifles.

18. MODEL 1890, SPANISH MAUSER

According to the records of Ludwig Loewe & Company the prime manufacturers, Spain adopted in 1890 a slightly altered Turkish pattern rifle. This rifle also has the box magazine projecting below the stock.

In 1892, 1893, and 1896, minor modifications in design were adopted by Spain. The only major change was the magazine which was made flush with the bottom of the stock, and was staggered.

These Mausers differ from the Belgian already described only in the following general details:

1. Caliber. The Spanish caliber is 7 mm or .276 inch. (Modern "7 mm Mauser" commercial ammunition can be used).

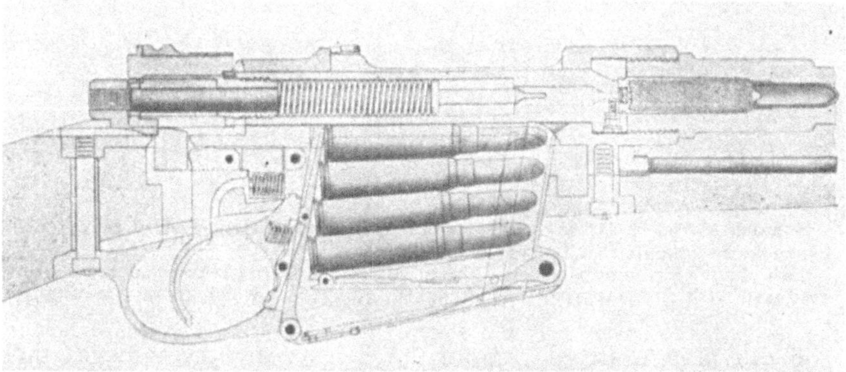

MODEL 90. SPANISH RIGHT SIDE VIEW WITH CHAMBER AND MAGAZINE LOADED

The cartridge is rimless and measures 3.08 inches overall and weighs 377.4 grains. The bullet is round-nosed and made of lead with a steel jacket. The bullet length is 1.21 inches, the maximum diameter is .284 inch, and the bullet weight is 172.8 grains. The charge is 38.3 grains of nitro-cellulose. The muzzle velocity is aproximately 2290 feet per second and the chamber pressure about 45000 pounds per square inch. (Modern specifications vary somewhat).

2. The Spanish pattern as used in 1892 is practically identical with the Boer rifle of the same year which uses the same cartridge.

3. Barrel is polished and browned on the outside and is turned down in steps with the portions between tapering. Sights are attached to barrel.

4. Rear sight has a thin tube underneath which encircles barrel and

is screwed and soldered to it. The right angle to the leaf is graduated for 400 feet. The face of the leaf itself is graduated from 400 to 2000 meters.

5. A half length rod for cleaning is provided. One end is screwed into a notch in the stock and the other end is provided with a slot for the cleaning rag. The rod is intended to be joined with others to form a common rod for several soldiers.

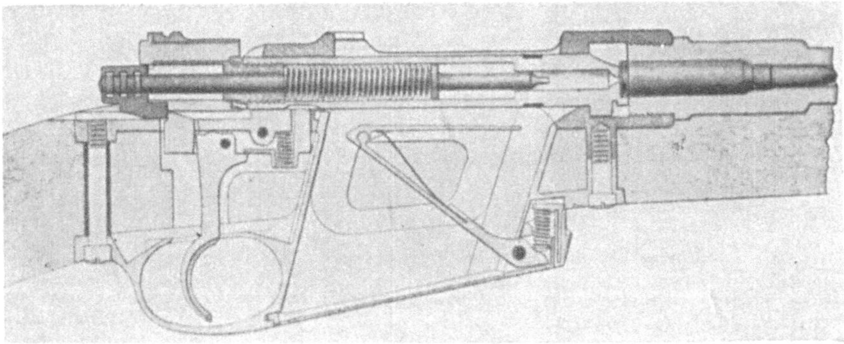

Spanish Model 92-93 Magazine Carbine, Cal. 7 mm. Arm cocked ready to fire cartridge in chamber, magazine empty.
This model has a straight-line single column box magazine, and must not be confused with the staggered box Model 93, which is a much more common arm.

6. All other differences from the Belgian are the same as those already given for the Turkish rifle.

Spanish Carbine

The carbine of this pattern is the same as the rifle with minor differences as follows:

The rear sight leaf is graduated from 400 to only 1400 meters. Bolt lever instead of projecting in straight line is turned down to lie close to the stock.

Barrel projects $\frac{1}{16}$ of an inch through the nose cap. The nose cap is extended upwards at its sides to form wings to protect the front sight. It is not equipped to receive a bayonet.

The lower band has a sling swivel on the right hand side. The rear attachment point for the sling is a plate below the rear of the small of the stock. The ring for the band is attached to this plate.

The weight, length and barrel length are correspondingly shorter.

The forestock comes to the muzzle on the carbine.

The carbine weighs about 7.5 pounds and measures 37-inches overall.

19. MODEL 1893 SPANISH MAUSER

In 1893 a very slightly modified version of this rifle was manufactured by Loewe for the Spaniards and shipped to Cuba. 30,000 rifles and carbines were delivered to Spanish troops there. These rifles were the backbone of the Spanish defense in Cuba during the Spanish-American War.

At the battle of San Juan Hill in Cuba, 15,000 of our troops attacked the hill defended by a mere 700 Spaniards, armed with Model 1893 Mauser rifles. The fact that the cartridges were smokeless, that the rifles were rapidly loaded with Mauser clips, and that in velocity and range they were far superior to both our single shot Springfield .45-70's and our Krags produced results which could not be overlooked.

At the outbreak of the War our Regular Army of 27,000 was equipped with .30 caliber rifles on the Krag-Jorgensen system. The Navy had purchased 10,000 Lee straight pull rifles of 6mm (.236) caliber. Most of the volunteers were armed with .45 caliber black powder rifles.

With only 700 men defending the hill, the Spaniards actually inflicted over 1400 casualties on our attacking forces! Twenty-one thousand one hundred fifty-four of these rifles and carbines were finally seized by our forces after the fall of the Spaniards.

After experimenting with numerous designs, we finally developed and adopted our famous Springfield Rifle of 1903, a Mauser pattern rifle better than any Mauser made before or since. With the adoption of the M 1 (Garand) semi-automatic rifle in 1936 we again forged ahead of the Germans in the field of rifle development for our armed forces; though for sniping and sharpshooting the Springfield continues to be the last word in military rifle design.

Other Mauser Rifles Based on the Spanish Model 93 Pattern

The following Mauser rifles are merely minor modifications of the Spanish Model 93:

The Turkish M93, the Swedish M94 (Caliber 6.5 mm), the Brazilian Model 94, the Swedish Infantry Rifle Model 96, the Chilean Model 95, the Uruguayan, Peruvian, Chinese, Transvaal, and Orange Free State (all Models of 95), and the Serbian Model 99.

Further slight modifications of these types have been made since the original issue. In general, however, the designs are the same.

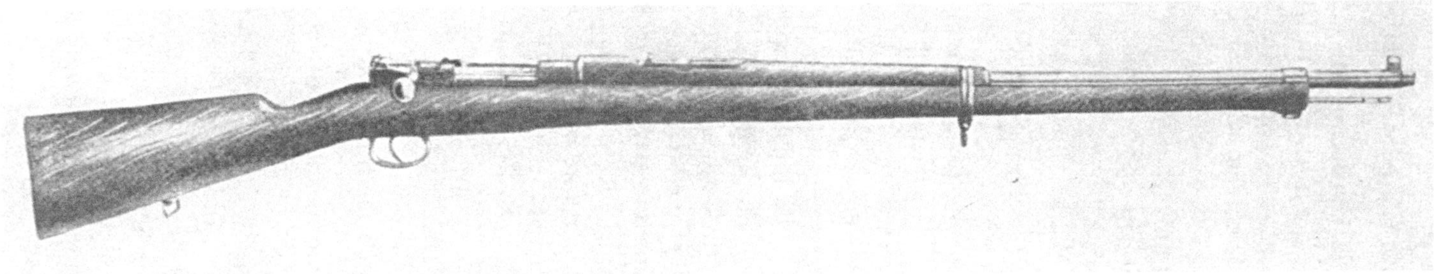

MODEL 93. SPANISH INFANTRY RIFLE. RIGHT SIDE VIEW OF ACTION

The receiver metal on the right side of the bolt opening has been cut away to facilitate ejection. This is a feature of all later Mauser rifles. The introduction of this rifle marked a great advance in the field of bolt action rifle design.

Modern 7 mm Mauser ammunition can be used. Bullet types and weights, as well as powder charges, differ with place and time of manufacture, but all will interchange. The 7 mm is commonly called ".276" in the U. S., but that described in European catalogs as "7 mm Mauser (.275)" will interchange.

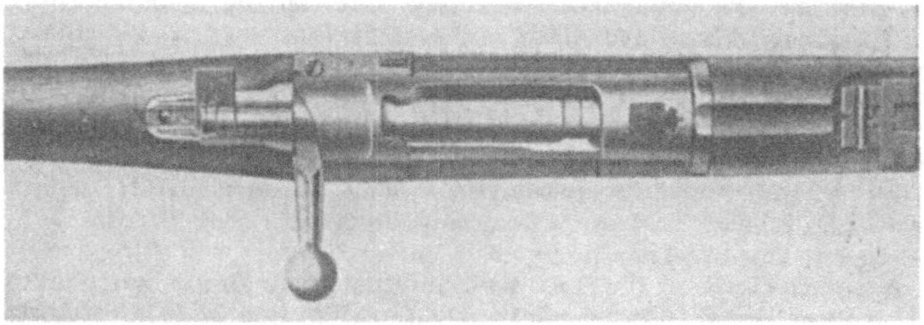

MODEL 93. SPANISH MAUSER. TOP VIEW OF ACTION

Note the improved style of bolt removal. This has been carried through on all later Mauser rifles. When the bolt is in full rear position, pushing out on the pivoted thumb piece seen at the rear left of the receiver, permits withdrawal of the bolt as a unit to the rear.

The camming surfaces on the top of the receiver show clearly how upward movement of the bolt handle compels the turning and withdrawing movement essential to effective primary extraction.

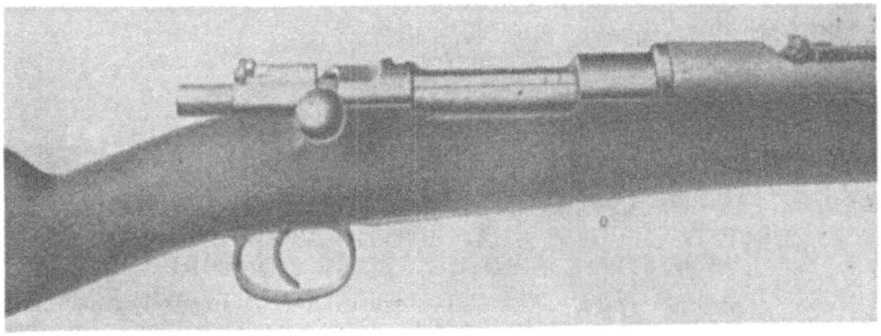

MODEL 93. SPANISH MAUSER. RIGHT SIDE VIEW OF RIFLE WITH ACTION CLOSED AND COCKED

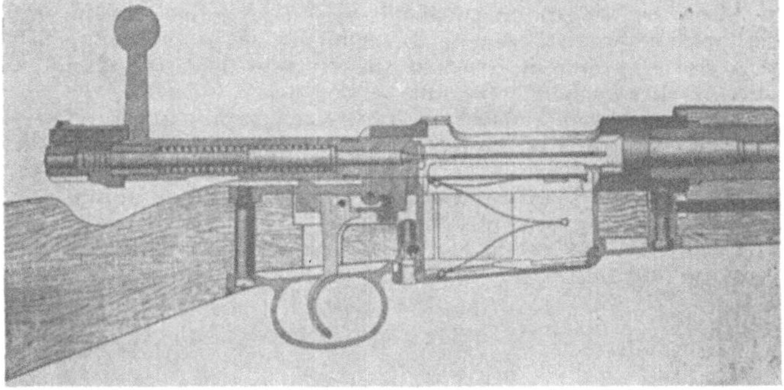

MODEL 93. SPANISH INFANTRY RIFLE. PHANTOM VIEW OF RIGHT SIDE OF ACTION SHOWING BOLT WITHDRAWN AND RIFLE READY FOR LOADING

A comparison of this drawing with that of the Spanish

Model 91, and Model 92-93, will show the principal differences in design. The magazine follower and spring construction and the magazine base plate removal system vary in all these models. The sear and trigger system of the 92-93 and 93 Models are alike and are improved over the Model 91.

Note that in this type of design while the initial opening movement of the bolt withdraws the point of the firing pin back within the bolt head, the spring is only partly compressed. Lifting and then thrusting down the bolt handle will compress the striker spring and cock the weapon if desired.

A comparison of the bolt and cocking-piece shape with that of earlier models will show graphically the alterations in design. This arm in 7 mm Spanish Caliber is the basis for the rifle designs of many South American nations. This rifle was manufactured not only by Mauser and by Ludwig Loewe & Co., but also at the Oviedo Arsenal in Spain and by Steyr in Austria.

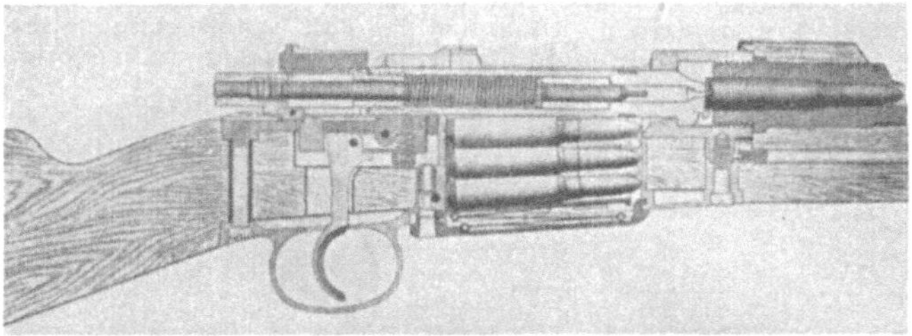

MODEL 93. SPANISH INFANTRY RIFLE. RIGHT SIDE PHANTOM VIEW SHOWING ACTION CLOSED, MAGAZINE AND CHAMBER LOADED

This drawing shows the magazine *fully* loaded and an *additional* cartridge in the firing chamber.

In rifles of this design, after a loaded clip has been inserted in the magazine guide and the 5-cartridges stripped down into the magazine, the top cartridge may be thrust further down into the magazine with the thumb of the left hand, while the bolt is eased forward with the right hand, until it clears the top of the cartridge in the magazine and is past the point where it can feed from the magazine. A sixth cartridge may then be inserted directly into the firing chamber and the bolt pushed home and turned down to lock.

Note that this rifle does *not* have a magazine cut off. Therefore, at the end of each rearward stroke of the bolt, the magazine spring brings a cartridge up in line for feeding. If it is desired to close the bolt on the chamber and retain the magazine loaded, it is necessary to manually thrust the cartridges in the magazine down and ease the bolt forward over the top cartridge.

20. GEWEHR 98. KAR. 98K. KAR. 98K42

(Note: *Mechanically these are practically identical, as are sporting Mausers in general*).

The German army, which had been using the Gewehr 88, adopted on April 5, 1898 the improved form of the Mauser rifle listed as "Gewehr 98." This arm was also introduced in short length as a carbine. In 1905 these rifles were bored to give larger groove diameter for the new "S" bullet.

In 1908 a further modification was introduced which was patterned after our 1903 Springfield to combine the features of rifle and carbine. Its length was intermediate between the two earlier forms. This was officially listed as the "Kar. 98." This type was widely used by the Germans in World War I; and again with only minor modifications it served in World War II as the "Kar. 98 K," and "Kar. 98 K 42."

The original "Gewehr 98" had a heavy rear sight. When issued with a somewhat lighter rear sight, it was classed as "Gewehr 98 A." These rifles weigh 9.5 pounds, measure 49.25 inches overall, and have a barrel length of 29.15 inches. Gew. 98 was the primary World War I German rifle.

The caliber is the standard German 7.9 mm but uses a new form of pointed bullet with higher velocity and generally improved ballistics. (When originally introduced, however, this arm used the standard M. 1888 cartridge. Adoption of a pointed bullet required increasing bullet diameter and groove diameter).

It is rifled with 4-grooves of .0065 inch depth, concentric, with a twist of 1 turn in 9.39 inches, to the right. Sights are the standard barleycorn front and the leaf rear graduated from 200 to 2000 meters.

When the Spitzer (pointed) bullet was introduced the ballistic requirements of the new bullet, which was shorter than the round nosed, required an increase in bullet diameter from about .318 inches to about .325 inches. Representative ballistics will be found herein.

MODEL 98. GERMAN INFANTRY RIFLE. RIGHT SIDE VIEW WITH ACTION CLOSED

This rifle is the prototype of all modern Mauser rifles. It introduced the feature of an extra or third lug at the rear of the bolt cylinder. This lug engages in a recess in the cylindrical part of the rear of the receiver to act as a safety factor in the unlikely event of the two locking lugs behind the cartridge case head letting go.

Another added feature of this design, and of the commercial 1904 and later pattern rifles by Mauser, is a small rib on the right side of the bolt cylinder which serves as a guide in withdrawing the bolt. When the bolt is closed this rib lies underneath and supports the extractor. The bolt face is recessed to receive the base of the rimless cartridge.

The arm as shown has the full 30-inch barrel length of the true rifle, the Gewehr 98. This arm was also made as a *short rifle* with a 24-inch barrel, and also as a true *carbine* with an 18-inch barrel.

The most modern Mauser rifles manufactured differ from this type only in such details as length, sights, furniture, and such modifications as were desirable for mass manufacture from stamped parts during war conditions.

World War II Mauser and Mauser System rifles, as well as all Sporting Mausers, are derived directly from this rifle. Mechanical differences are inconsequential. This is a basic design.

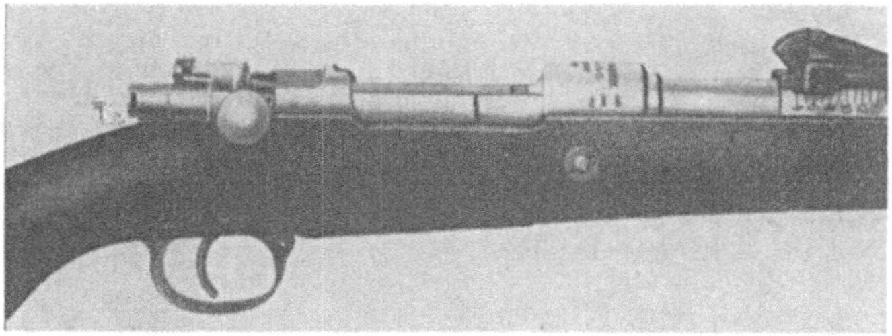

MODEL 98. GERMAN INFANTRY RIFLE. RIGHT SIDE VIEW OF RECEIVER WITH ACTION CLOSED

While this weapon as shown bears the trade mark of Waffenfarbik Mauser, the arm may also be found stamped with the names of the various German manufacturing arsenals.

MODEL 98. GERMAN INFANTRY RIFLE. RIGHT SIDE VIEW OF RECEIVER WITH ACTION OPEN READY FOR LOADING

Note the third locking lug directly ahead of and below the bolt handle and also the long rib bolt-guide on the bolt cylinder in line with the bolt handle. These are special features found in all late model Mauser designs.

The date 1907 is year of manufacture. The bolt bears the receiver serial numbers. Other parts bear the final 2 digits, showing original factory checking and assembly.

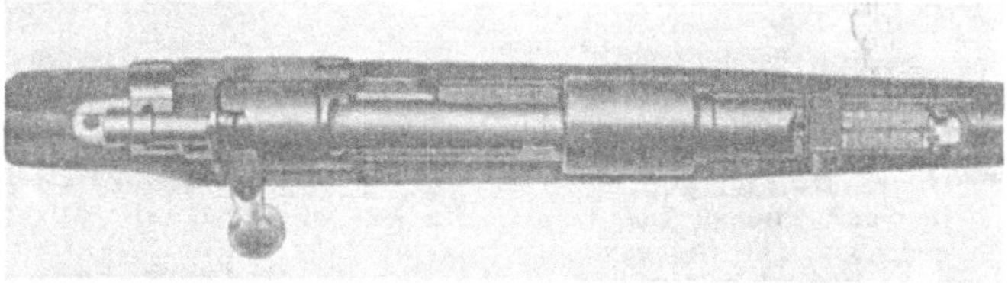

MODEL 98. GERMAN INFANTRY KAR. TOP VIEW WITH ACTION CLOSED

The cutaway section at the front of the metal receiver bridge across the rear of the action provides a guide for insertion of the loaded clip. As the bolt handle is turned up, its lower end where it faces the cylindrical bridge section of the receiver moves against the inclined plane to provide the leverage necessary to loosen the cartridge in the chamber and provide primary extraction. Thus as the handle is raised, its bearing against the plane (or cam) surface on the receiver bridge compels the unlocked bolt to pull back away from the face of the cartridge chamber.

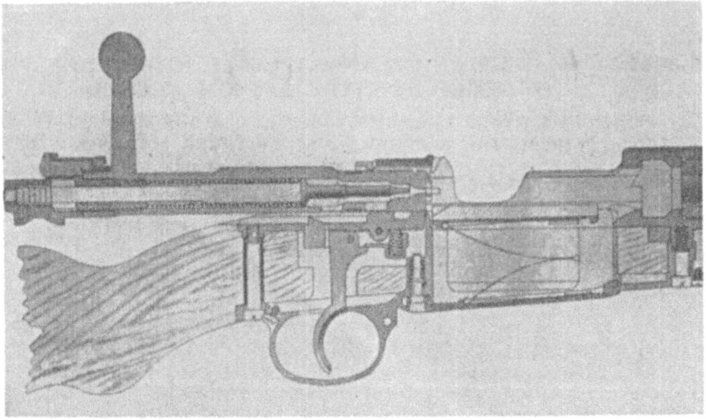

MODEL 98. GERMAN INFANTRY RIFLE. RIGHT SIDE PHANTOM VIEW SHOWING ACTION OPEN READY FOR INSERTION OF LOADED CLIP

Note that the left side of the receiver above the magazine way is cutaway. This is to provide clearance for the thumb as it pushes cartridges down off the clip into the magazine.

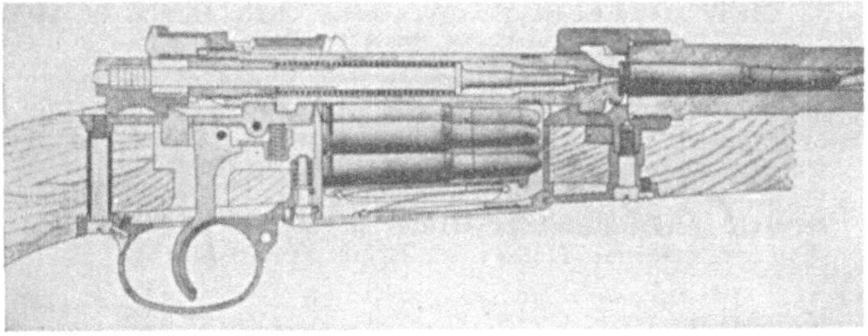

MODEL 98. GERMAN INFANTRY RIFLE. RIGHT SIDE PHANTOM VIEW WITH ARM READY TO FIRE

Four cartridges remain in the magazine. After the cartridges are stripped down out of the clip, thrusting the bolt forward knocks the empty clip out of the guide lips and chambers the top cartridge, leaving four in the magazine.

As already described in the Spanish Model 93, by easing the bolt forward over the top cartridge while holding the magazine cartridges thrust down with the thumb, it is possible in this design to insert a sixth cartridge in the chamber. If loading is done directly by the bolt however, only 5 can be inserted.

Note the third (safety) lug locked into place to rear of magazine well.

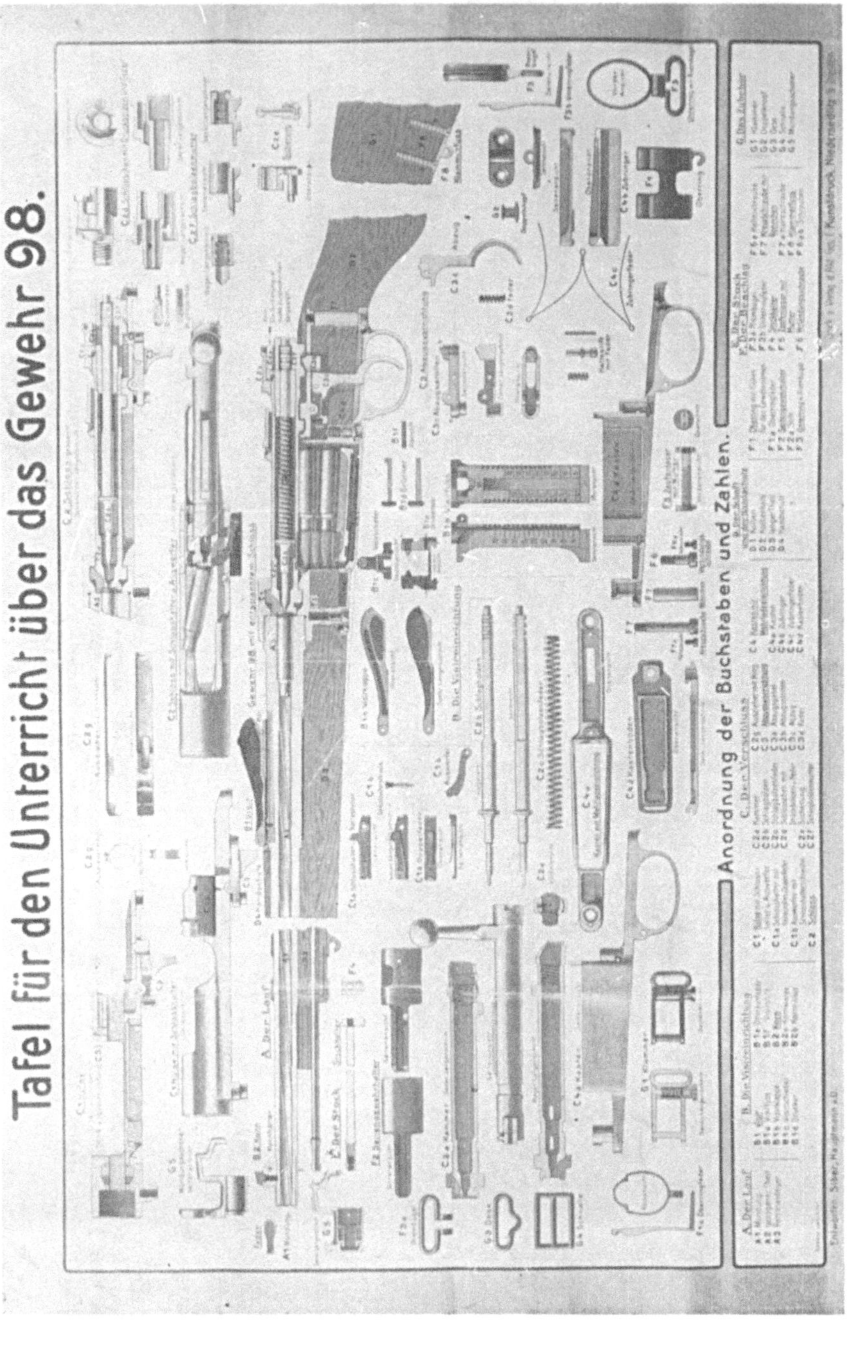

German ordnance chart showing all details of Gewehr 98. The mechanism of this rifle is similar to all later Mausers. While some of the parts are stamped and others forged, mechanically they are practically identical. Description of basic Mauser design on pages 121-128 essentially covers all Mausers made since 1898.

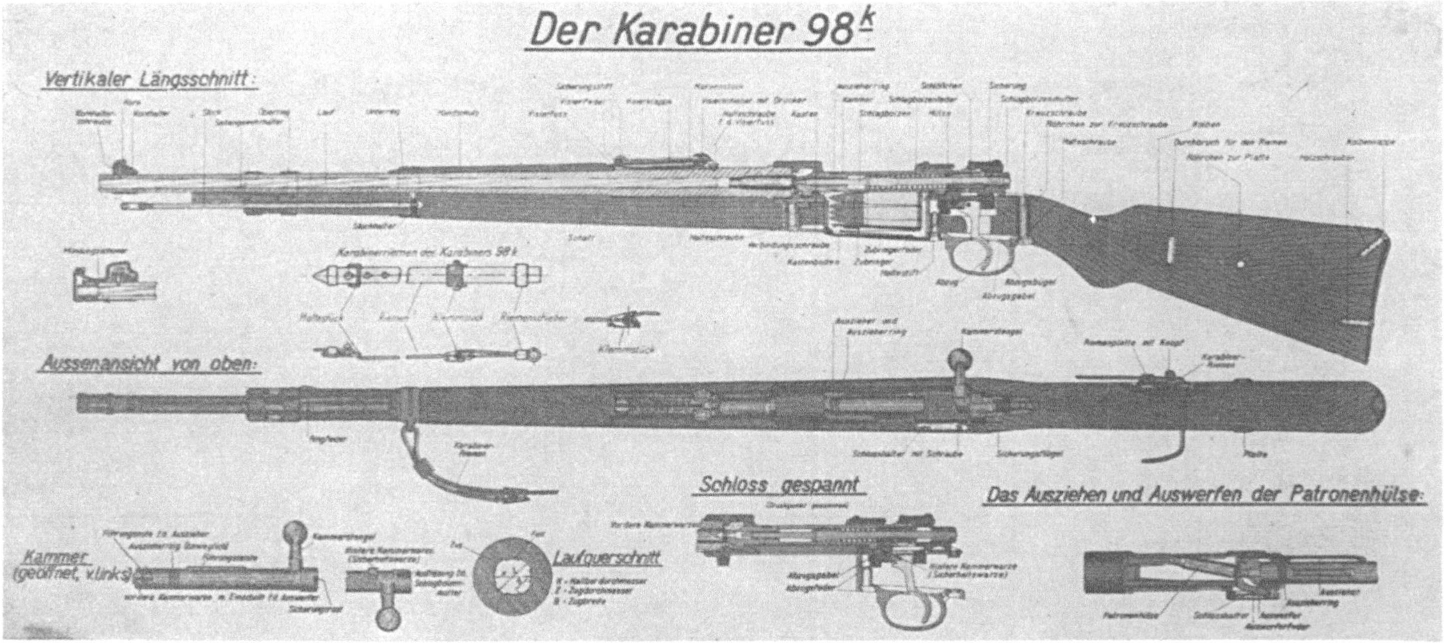

German ordnance chart of the last Mauser model. Mechanically it is the same as the Gewehr 98.

All Mauser and Mauser System rifles wherever made since 1924 are based on the above design. This differs from the World War I 1898 design in having a broader gas flange; a magazine follower which holds the bolt open when last shell is ejected, different sights and stock. It is also shorter. Mechanically all are nearly indistinguishable.

The Evolution of the Kar. 98

The Kar. 98 underwent no change of any kind from 1908 to 1916. In 1916 a hole was bored through the stock and provided with a bushing. This hole which has been characteristic of all Mausers of military type since that time, serves several purposes. It was originally designed to permit passing a locking rod through the sides of a rifle case, and through the holes in the butts to secure cased rifles for shipment and storage. It was later used as a resting point for the striker in dismountin the bolt to prevent injury to the point. During World War I it served as a special mount for a trench periscope.

In 1907 finger slots were provided below the rear sight to permit a firmer grasp with the fingers.

During World War I various models of sniping rifles were made, many using 20-shot box magazines, and some equipped with bolt covers. The essential mechanical design, however, has not changed; and most important operating parts are interchangeable.

The "Kar. 98," the "Kar. 98 K," and the "Kar. 98 K 42" are practically identical. They differ only in minor points of construction or in manufacturing processes.

They weigh about 9.5 pounds, measure about 43.5 inches overall, and have barrels of about 23.4 inch. Rifling is the same as in the Gewehr 98.

The foresight may be blade or barleycorn with a V-notch radial rearsight.

The descriptive and operational data on pages 121-138 substantially cover all Mauser models wherever made after 1898. Specifically that data is based, however, on the Mauser 1924 series; the pattern was evolved from the Gew. 98 after World War I to increase efficiency and permit mass manufacture.

Belgium, Chinese, Czech, Polish, Portuguese, Spanish and Yugoslav "Mauser System" manufacture is practically identical with the German.

21. MISCELLANEOUS EARLY MAUSER RIFLES, 1895 TO 1907

In 1895, 1896, 1900, 1903, 1904, 1905, and 1907, various minor modifications of Mausers were made at Oberndorf, for various foreign government. Those after 1904 differ essentially only in caliber.

During the period the Germans made wide manufacturing contacts. Arrangements were made for the manufacture of their rifles at the great F. N. plants at Herstal, in Belgium, and this firm later filled export orders when Germany was not permitted to do so.

During this period also the Mauser and the Mannlicher plants tooled up to manufacture specimens of each others weapons at various times. Many 1907 model Mausers for the Mexican Government in caliber 7 mm, were manufactured at the Steyr works in Austria.

Spain has built rifles for her own forces and for South American export on all Mauser Systems since 1893.

The degree of interchangeability in parts in these weapons is surprising. All rifles manufactured under Mauser patents from 1893 to 1907 utilize the same bolt heads; and the bases of the respective cartridges are the same. Thus it was necessary only to make a barrel change, and a minor magazine change, in order to transform from one caliber to a larger one.

Furthermore, the case lengths of the various cartridges developed are so nearly alike that very little receiver work is necessary to convert the manufacturing machinery from an "export" caliber to one which will handle the standard German cartridge.

In 1924 and later, various European factories were tooled up to manufacture the Mauser-System rifles as modified in Germany on the Gew. 98 design. These are the same except for minor details as the Kar. 98 K.

When the Germans overran Austria, Belgium, Czechoslovakia, and Poland each in its turn, they were able to utilize much of the equipment they seized since it was originally adapted to German ammunition; and were further able to put seized plants to work manufacturing arms and ammunition for their standard weapons with a minimum of difficulty. In the years of peace they had seen to it that the factories in those countries, through cooperation and subsidy (among other means) were equipped with machinery and designs of German development and selection.

DESCRIPTION OF MODERN MAUSER RIFLES, 1898-1944

All German service rifles manufactured since 1898 and all genuine Mauser sporting high power rifles manufactured since 1904 are mechanically practically identical.

Types from 1924 on have improved gas flanges. Sights and stocks are improved. Barrels are shorter. However, basic design is the same.

Some varieties of the sporting Mausers were made with set triggers, but otherwise, are mechanically identical with the army type.

The receivers are specially heat treated in all models and in peace time were tested to a chamber pressure of 100,000 pounds per square inch, or practically double the maximum pressure produced in standard army rifles of German or United States design.

Headspace and Gas Escape

Headspace on genuine Mausers is always held to very close tolerances. The cartridge case head protrudes about 2.8 mm past the face of the chamber into the lug well when seated with the action open. The barrel and the recessed face of the bolt surround the cartridge completely on all sides except at the extractor cut, thereby reducing the danger of blown cartridge case heads to an absolute minimum.

A ring shaped groove machined into the receiver between the chamber and the locking lug, is provided to receive the cylindrical bolt head to make a particularly close fit. Any amount of gas which might eventually get through this opening past the bolt head is allowed to escape through openings in the bolt which prevents the flame from spurting to the rear and injuring the shooter.

Two gas vents in the front part of the Mauser bolt permit the safe escape of gas in the event of a blown or punctured primer. The vents are in front of the firing pin plates which prevent the gases from reaching the spring. They are lateral slits 10 x 5 mm (.39 x .195 inch). These are comparatively large, and are so designed intentionally to permit immediate diffusion of any escaping gas.

Firing Pin Design

Another feature of Mauser design is the one-piece firing pin of very heavy weight. The firing pin has a safety in the form of a band for the

THE GERMAN WORLD WAR II KAR 98K

1. With laminated stock. 2. With plastic stock. 3. With special winter trigger.

Mechanically these are all practically the same as the Gew. 98. However, they have the wide gas flange, magazine follower bolt stop, and improved sights, etc. developed in 1924.

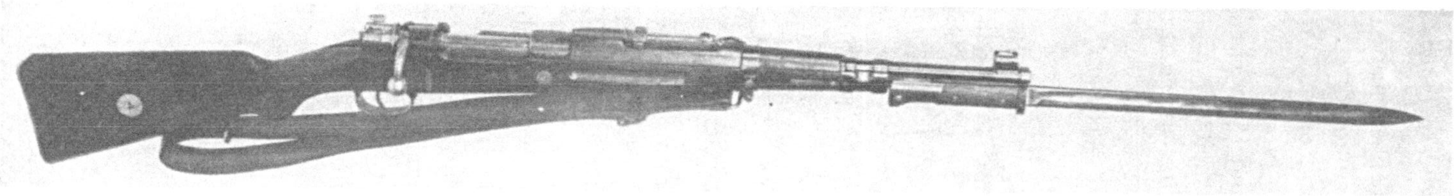

Persian Mauser Carbine. Caliber 7.92 mm German.

Mechanically the same as the German. Belgium made these "Mauser System" rifles and carbines for general export. Barrel lengths, stock types, slings, sights and calibers could be dictated by the purchaser. F. N. (Belgian) markings and/or native markings were left to discretion of purchasing nation.

spring to rest upon, having two flanges diametrically opposed. When the bolt is locked these flanges fit into corresponding grooves on the inside of the bolt. When the lock is opened, however, or only partly opened, the position of the groove prevents the firing pin from snapping forward of its own accord under any circumstances.

The Mauser bolt plug (called "bolt sleeve" on our Springfield) is made oversized so that a projecting rim on the forward part, covers the rear end of the bolt as a positive stop against any rearward flow of gas. This is commonly called a "gas flange."

Magazine

The magazine and triggerguard are made in one piece. The magazine base is so constructed that it may be released on its hinge from below by pressure on the button in front of the triggerguard. This permits unloading the magazine without operating the bolt and functioning the cartridges through the mechanism. This magazine is closed at the bottom with a plate flush with the stock. The capacity is five cartridges, held in double staggered rows.

Bolt Characteristics

The bolts on these rifles differ from those of earlier Mausers, and of most other bolts, in that a third (or safety) lug is placed on the rear section of the bolt, and that a travel-guide rib is also provided.

The bolt itself is of very simple and strong construction. It is one piece with the bolt head which is *not* removable. Two locking lugs are placed opposite each other at the front end to lock into corresponding recesses in the receiver. The third or safety lug at the rear of the bolt engages in a recess in the cylindrical part of the rear end of the receiver. In the unlikely event that both the front lugs should let go, this rear lug is designed to hold the bolt.

The bolt handle or lever may stand out straight at right angles as in the early military patterns, may be turned down as in the later Kar. models, or may be specially shaped to lie flat against the stock as in some of the sporting types.

A cam shaped recess at the rear end of the bolt receives the stud of the cocking-piece. Thus when the bolt handle is turned up, the camming action withdraws the striker from the face of the bolt, since the cocking piece is so shaped that it cannot turn and hence is forced straight back together with the striker. At the opposite side there is

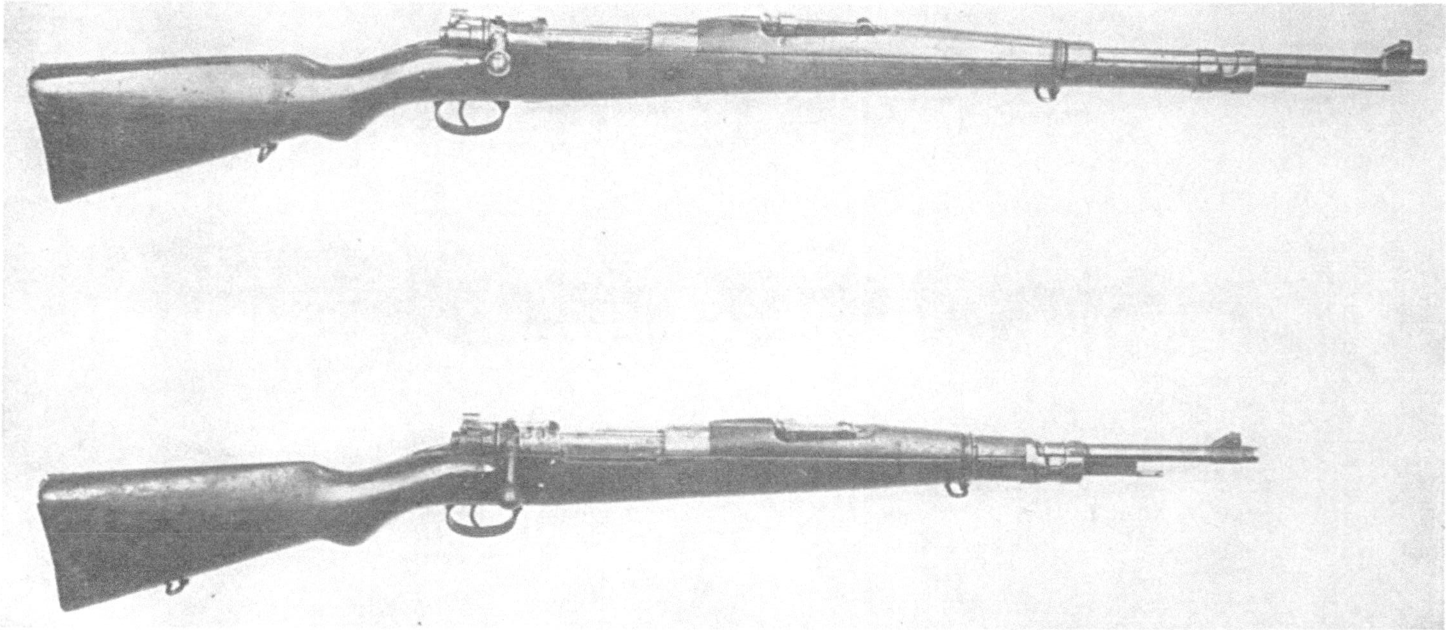

Belgian (FN) Mauser System. Cal. 7.92 mm. Rifle and Carbine with Ethiopian markings.

All F. N. rifles of this pattern regardless of country for which they were built, have the mechanical characteristics and essential measurements of the German Kar. 98 k receiver. Models may be dated 1924, 1924/30, 1935, etc. Calibers are usually 7, 7.65 or 7.9 mm.

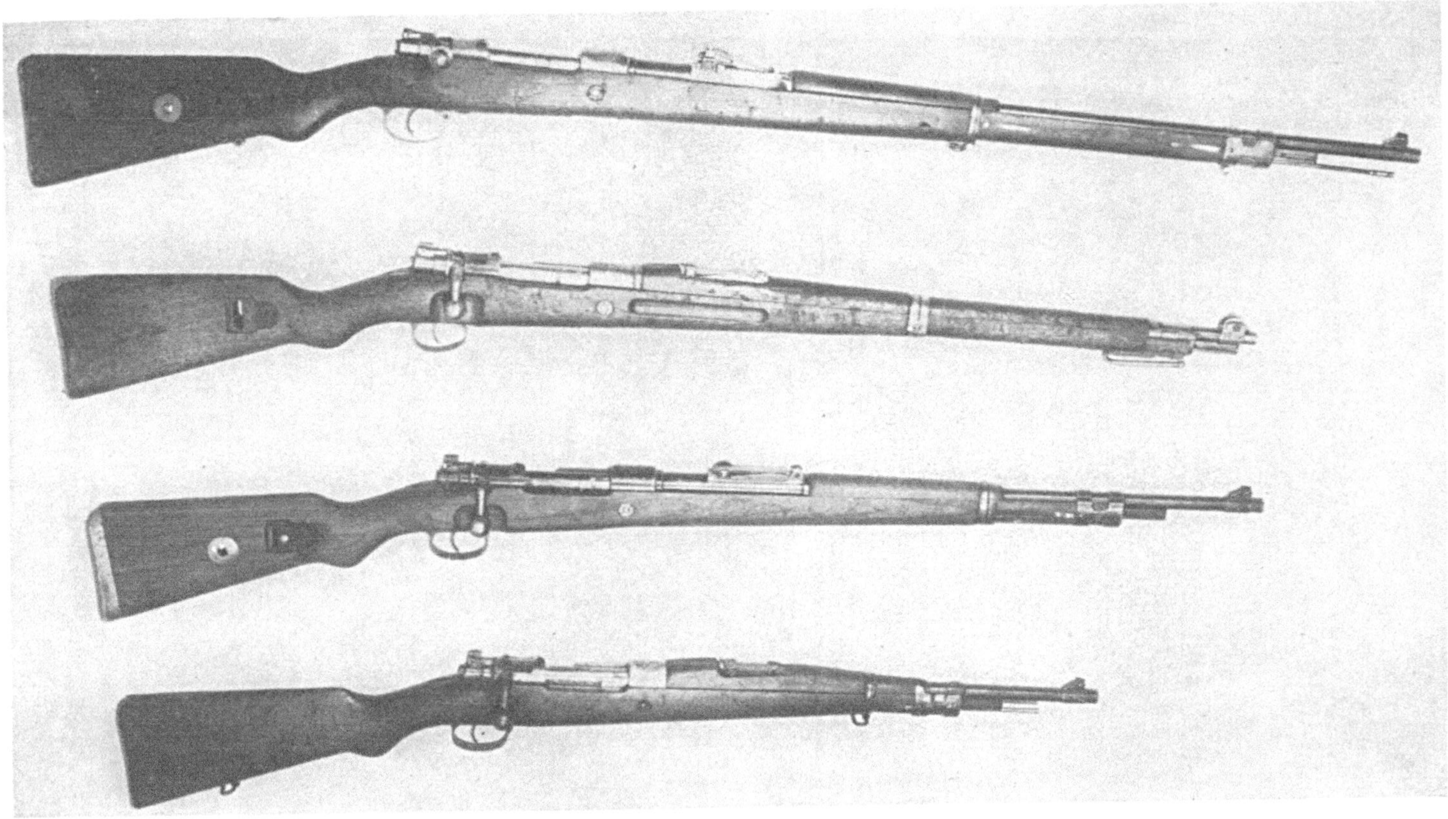

Top to bottom: 1. German Gewehr 98. 2. Carbine 98. 3. Pattern 1924 Kar. 4. German carbine.
All are based on the Gew. 98 System. Pattern 1924 is slightly longer than late Kar. 98k but is otherwise much the same.

a small notch to receive the tooth of the safety bolt. A small rib on the bolt body acts as a guide in withdrawing the bolt. When the bolt is closed, this rib lies below and supports the extractor. The face of the bolt is specially recessed to take the base of the rimless cartridge which it encloses entirely except at the extractor point.

The one-piece striker has a short point, and a collar against which the mainspring bears for cocking. The rear of the striker has three interrupted grooves to connect it with the cocking piece. The rear portion of the pin is flattened on two opposite sides to keep it from revolving inside the bolt plug.

The mainspring when not compressed measures about 5 inches long, and is composed of 28 to 31 coils of .06 inch diameter wire, the number of coils varying with the pattern of the rifle.

Interrupted lugs on the cocking piece permit it to be connected to the striker pin by giving it a quarter turn. Note that it is impossible to join these incorrectly, since the rear groove on the striker and a corresponding bearing in the cocking piece, are broader than the two front ones.

A projection on the under side of the cocking piece travels in a groove cut for it in the tongue in the receiver. When the bolt handle is turned down, this projection is engaged with the sear nose so that the bolt handle turning down the inclined face of the receiver bridge pushes back the stud to compress the mainspring. The bolt lugs passing along the cam slots to their locking recesses add to the mainspring compression thrust. The forward top surface of the stud is chamfered to mate with the sloping surface at the end of the bolt; thus on turning up the bolt, the cam surface forces the stud back to give initial compression to the spring and also withdraw the point of the striker from the fired primer.

The bolt plug (sleeve) screws loosely into the rear end of the bolt by means of a reinforced thread. It closes the opening through which the striker is inserted when assembling, and also serves as a seat for the mainspring. The bolt plug does *not* turn when the bolt is operated as the cocking stud works in its slot. The front part of this bolt plug is semicircular in shape, providing a flange which acts as a shield to protect the user against accidental back blast in case of a blown cartridge case or primer.

On the left side of the bolt plug there is a pin catch which when in position engages with the bolt and holds the bolt plug and bolt se-

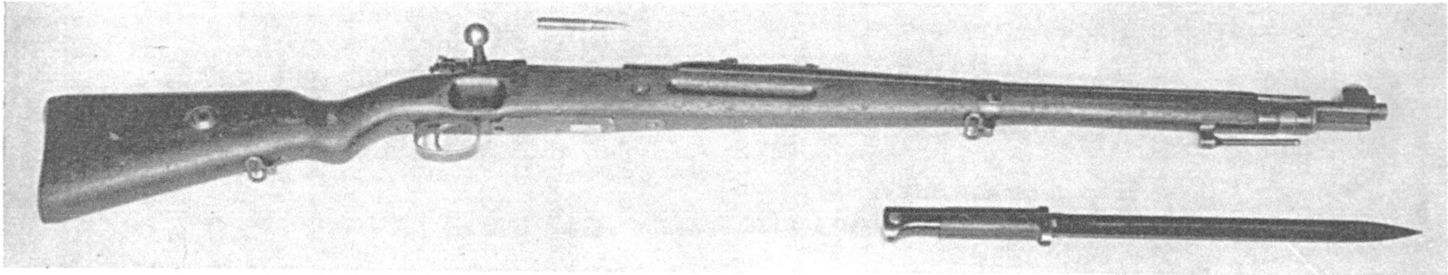

Polish Mauser Kar. 98. Cal. 7.92 mm.

Except for minor external features, this arm is the same as the German Kar. 98 of World War I. Some later models of this Polish manufacture also have the modified bolt and magazine of the Kar. 98k. Latest pattern Polish designs are even stocked like the Kar. 98k. Polish rifles were made at Warsaw and Radom arsenals.

Portugese Mauser. Cal. 7.92 mm. The latest official model.

This arm is practically identical with the German Kar. 98k. It is replacing the old Mauser-Verguiero (Portuguese Service M. 1904 rifle, caliber 6.5 mm).

curely locked against accidental unscrewing during opening or closing movements. A cylindrical hole is drilled in the top of the bolt plug to receive the bolt of the thumb safety which is positioned on the assembled firing pin.

The safety bolt itself consists of a thumb piece and a spindle which works in a hole in the bolt plug. When the thumb piece is turned to vertical position, and the bolt is closed, the flange on the safety bolt is thrust up in front of the top of the cocking-piece, forcing it back a little and withdrawing the stud from contact with the sear. This positively prevents release of the striker. When the thumbpiece is turned over to the right, the cocking-piece is still locked; but at the end of the stem where it is not cut away it enters into a slot on the end of the bolt, locking the bolt and bolt plug together, to prevent the bolt from being revolved.

The safety bolt is held in the two safety positions by the top of the cocking-piece bearing on shallow depressions on a flange in the safety bolt itself.

The Extractor

The extractor is a wide, long H-spring, terminating in a claw. The front of the claw is chamfered to enable it to ride over the base of the cartridge, as the cartridge is chambered, thereby enabling the hook to engage in the groove at the base of the cartridge, and support it solidly against the bolt head ready for extraction. An undercut groove on the extractor engages on dovetail projections on the spring band which revolves in a groove on the bolt. A projection working in a groove at the bolt head permits the pull of the bolt to be transmitted to the extractor.

When the bolt is drawn back the right hand lug lies under the extractor in a recess provided for it. This feature of a non-revolving extractor is particularly valuable because it prevents friction between the extractor and the face of the discharged cartridge, and also because it does away with the necessity of cutting away the rear end of the barrel to afford clearance for the extractor end.

The Receiver

The receiver, to which the barrel is screwed, is grooved on either side of the boltway to provide for the working of the bolt lugs and extractor. The right hand groove is partly cutaway to permit ejection

Belgian (FN) Mauser System. Cal. 7.92 mm. Model 1924. This is the common European and South American late model Mauser. It may be encountered in any of the standard Mauser military calibers, varying with the country for which it was made.

Basically this is the same as the latest German pattern. Models 1930, 1935, etc. are minor variants only.

Shape of bolt handle, length of barrel, stocks, slings and similar details were arranged to suit the purchaser of a large quantity order. As a result, dozens of minor variants are encountered, though all are built on the German Kar. 98 k type action.

German Gewehr 33-40. Cal. 7.92 mm. Czech modification.

Basically this is the same as the German Service Kar. 98k. Barrel was shortened and special steel side butt plate provided for use by special mountain troop units. Originally made by the Czechs in 1924 and slightly modified in 1930, it was used by Germany from 1940.

of the case to the right. The twin recesses for the locking lugs are at the forward end of the receiver and are connected, of course, with the travel grooves.

The rear section of the receiver through which the bolt must travel forms a complete cylinder (or bridge) in whose forward end the clip guides are machined; while to the rear of this it lengthens into a tang which is recessed to permit the cocking stud to move over it. This tang is bored and threaded to receive the rear action screw which aids in holding the action, triggerguard and stock together. An opening is provided in the tang to permit the sear tooth to project. A cross groove is cut directly behind the magazine way to receive the third safety locking lug.

The bottom of the receiver is cutaway to provide for the magazine, which is one piece with the triggerguard and which is fastened to receiver and stock by screws.

A ribbon spring secured to the magazine bottom plate and the underside of the follower serves to thrust the follower up and support the cartridges above it. The follower has a raised rib on its left side which elevates the cartridge column above it almost to the level of the right hand column, and thereby serves to bring a cartridge up from each side alternately to be held by the overhanging magazine lips until stripped forward by the bolt for chambering. A small stud to the rear of the magazine well serves to secure the magazine bottom plate. (In some sporting models this takes the form of a swinging lever below the bottom plate). This stud can be depressed by the point of a bullet and the plate can then be slid backwards and removed, bringing the spring and follower with it.

Magazine Design

Special box magazines holding additional numbers of cartridges up to 20, were provided for military purposes, to be inserted in place of the bottom plate of the standard magazine, but the standard holds 5.

The magazine may be loaded by individual insertion of cartridges, each one being pushed down into the magazine, or may be loaded by stripping in five cartridges directly from the clip. The clip itself, is the familiar thin sheet of steel, having edges turned over to engage in the grooves in the cartridge cases, while a spring of wavy ribbon steel, fastened by two small tangs to the inside of the clip, holds the cartridge heads firmly.

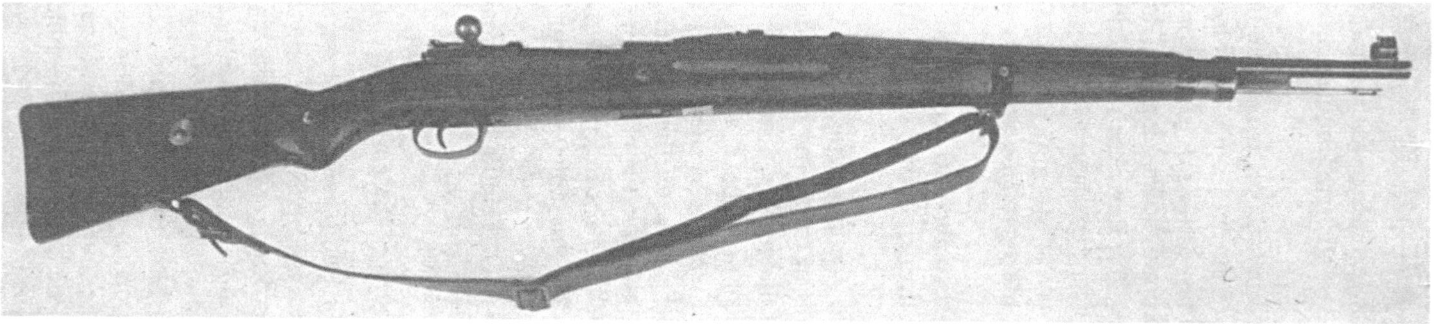

Czech Mauser VZ24. Cal. 7.92 mm.

This design was copied and adopted directly from the German improvement of the Gew. 98 in 1924. It was sold throughout the world in export trade at a time when Germany could not export military rifles.

This is one of the finest Mauser System rifles made.

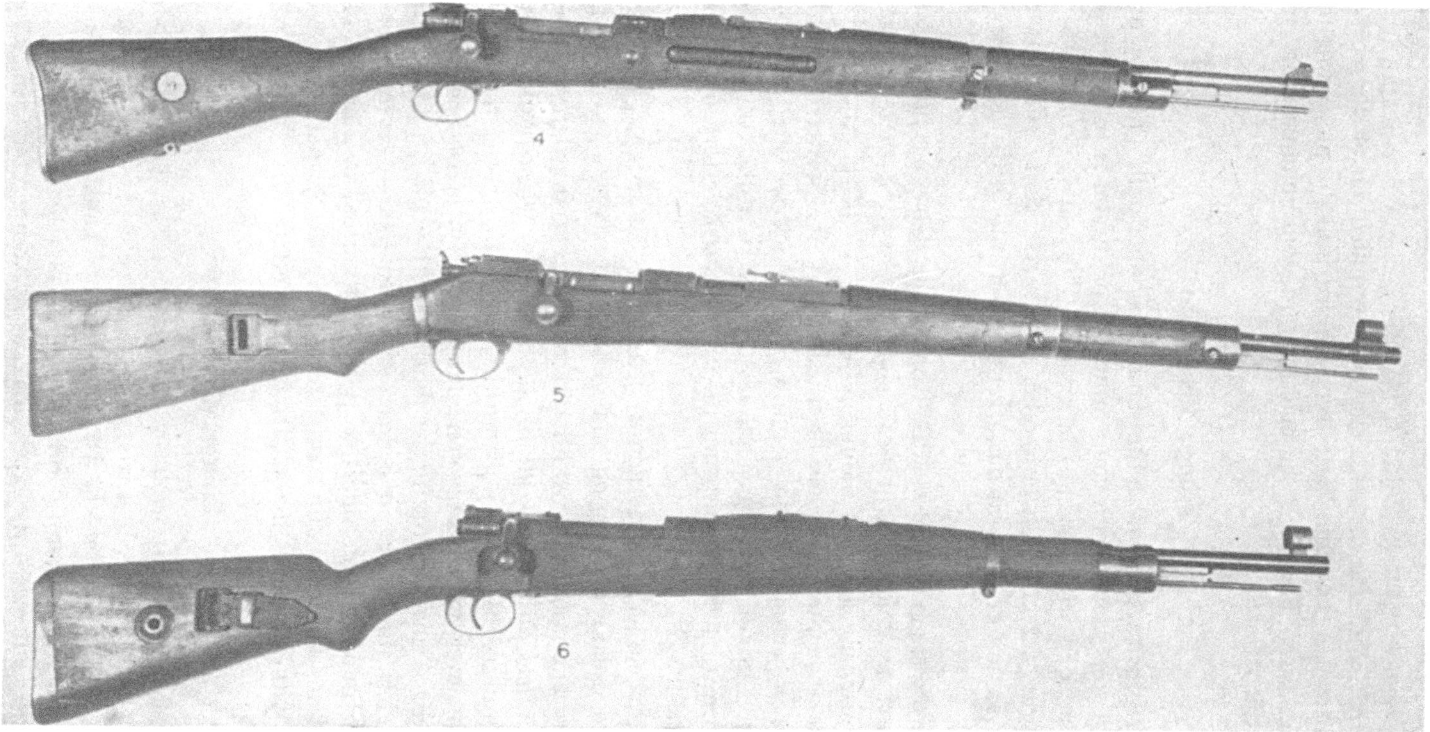

OTHER GERMAN SERVICE ARMS IN WORLD WAR II

4. Kar. 98K with modified stock. 5. Gewehr 98-40 (not a Mauser). This is a turning bolt modified Mannlicher-Hungarian type. Cal. 7.92 mm. 6. Gew. 33-40. Cal. 7.92 mm. Modified Czech Mauser System.

Note that while the cartridge clip may be pulled out of the receiver guide, after the cartridges have been stripped into the magazine, this is not necessary. Thrusting the bolt forward, will automatically eject it from the action.

Trigger Mechanism

The trigger mechanism is the standard double pull. The sear projects into the groove in the tang provided for it, and is part of a bar pivoted near its front end to the receiver. The sear is actuated by a spiral spring let into it, whose other end bears against the receiver. The bar section of this sear has a vertical slot in which the trigger is pivoted.

When the trigger is pressed, the front of the two projections on the upper surface of its arm, presses against the receiver, thereby lowering the sear, and giving a sliding movement. At this point, the leverage now that the slack has been taken up, is transferred to the rearmost projection. At this point, a slight check is noted, and continued pressure on the trigger releases the sear from the cocking-piece stud, permitting the mainspring to drive the striker forward to fire the cartridge.

Note that no half-cock system is included in this design. The striker is cocked usually by opening and reclosing the bolt.

To prevent accidental withdrawal of the bolt from the receiver, a stop hinged on the left side of the body is provided with a tooth which projects through a slot into the groove for the left lug of the bolt. This lever, which is pivoted on a pin, is kept pressed against the receiver by a small flat spring let into it. As the bolt is pulled to the rear, its left lug comes in contact with this tooth, preventing any rearward movement beyond that point. Pulling out on the bolt stop pivots it to withdraw the tooth from the path of the lug, which permits withdrawal of the bolt from the receiver.

Ejector Mechanism

The Mauser ejector pivots on the same pin as the bolt stop of which it forms a part. It is triangular in shape, and is flat. It is actuated by a spring inside the bolt stop lever. There is a slot cut in the left locking lug and in the face of the bolt head for passage of the ejector; and when the bolt is withdrawn, the ejector springs into

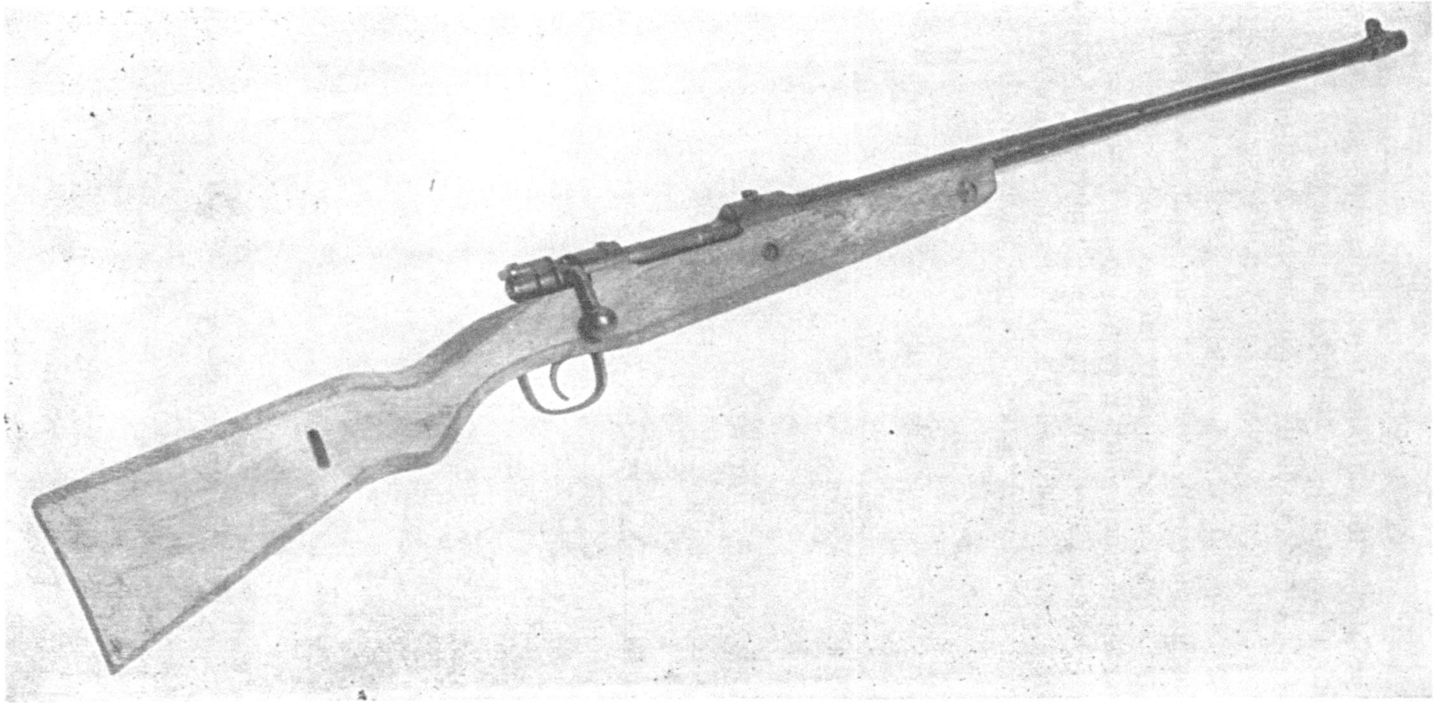

Mauser Kar. 98k. Action modified for volksturm use. Cal. 7.92 mm.

In the emergency, Germany barreled and stocked some Mauser actions as shown. Bolts as a rule have rough squared lugs. Assemblies are usually riveted. Some designs have detachable magazines. These were designed as highly expendible weapons. Care should be exercised in using them.

this slot. The base of the cartridge being gripped by the extractor strikes against the ejector. Since the extractor continues to draw the right side of the cartridge, and the ejector pressure is on the left, the cartridge or cartridge case is swung around and thrown out of the grip of the extractor and out of the action. The stock is in one piece, and on all models after 1904, is provided with a half pistol grip. The fore-end is attached to the barrel on the military model by two bands. A hand guard is provided which is secured to the lower band. The butt plates are of steel in most military types. In sporting types, they may be hard rubber, or cushioned. A half length cleaning rod is provided, which is threaded to permit being joined to another rod.

23. SEMIAUTOMATIC MAUSER RIFLES

Model 98. Mauser Semi-Automatic Rifle

The success of the original Mauser automatic pistol encouraged the great inventor to turn his developing talents to an attempt to create a military rifle in which mere pressure on the trigger would be sufficient to fire each succeeding shot.

The Model 98 was the first rifle of this sort produced by Mauser which proved at all practical. However, it was never in general production and will normally be encountered only in rifle collections.

The original German Reich patent No. 105619 covers the use of the turning lever or valve type of action designed by Mauser. Modifications of this system were used in a wide variety of later arms. It was not until its appearance in the Russian light machine gun, the Degtyarov which first saw service in the Civil War in Spain, that this locking principle was developed as a truly efficient mechanical design.

Later German designs, notably the semi-automatic Kar. 43 which the Germans employed in the closing years of World War II to some extent, also used a modification of this original system of operation. While this latter was a gas rifle, and the original Model 98 as introduced by Mauser was a short-recoil operated rifle, it should be noted that the locking principle is equally applicable to both operating systems. The lock is often called a "claw block" or "valve-type" lock.

This design uses the familiar recoiling barrel which is firmly locked to the breechblock at the instant of explosion. The barrel floats within an outer barrel or casing. It is supported at its muzzle end. In these early designs, locking levers were mounted in an extension of the barrel and when the action was fully forward their rear locking arms were securely engaged in cuts in the sides of the breechblock. After the short recoil, and as the barrel hit its stop and its travel was halted, the action had moved back far enough to permit camming action to thrust the rear of the pivoted bolt lugs in when they hit the necessary surface on the barrel extension, thereby drawing the rear locking surfaces out of engagement with the breechblock.

The breechblock then travelled back to extract and eject and compress its recoil spring in standard fashion.

The standard military type Mauser magazine holding its cartridges in staggered rows thrust a cartridge up in line to be stripped by the breechblock into the firing chamber on the return stroke when the

MODEL 98. MAUSER SEMI-AUTOMATIC RIFLE WITH CLAW BLOCK LOCK. SHORT RECOIL SYSTEM.

This rifle first patented in Germany on February 20, 1898 is the earliest of Mauser's self-loading rifles. No Mauser self-loading (automatic) rifles designs were ever made in quantity. Few went beyond the experimental stage.

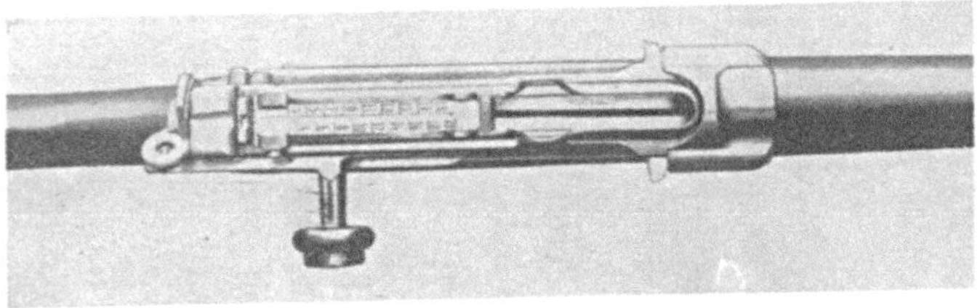

MODEL 98. MAUSER SEMI-AUTOMATIC RIFLE. TOP VIEW WITH ACTION OPEN.

A comparison of this photo with the closed photo will show the short stroke of the operating handle in this rifle design.

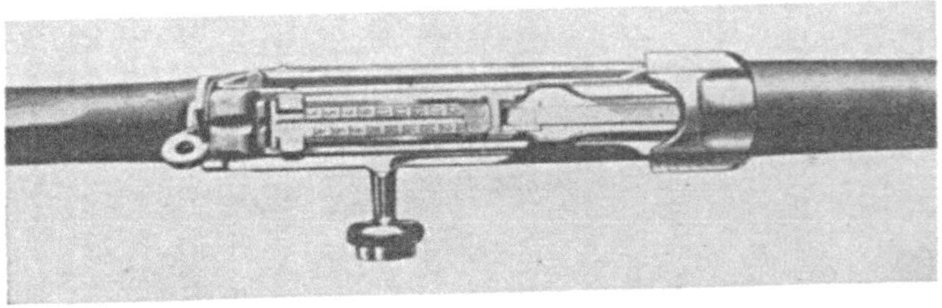

MODEL 98. MAUSER SEMI-AUTOMATIC RIFLE. TOP VIEW WITH ACTION CLOSED.

Note position of cocking handle and mounting of rear sight in this design.

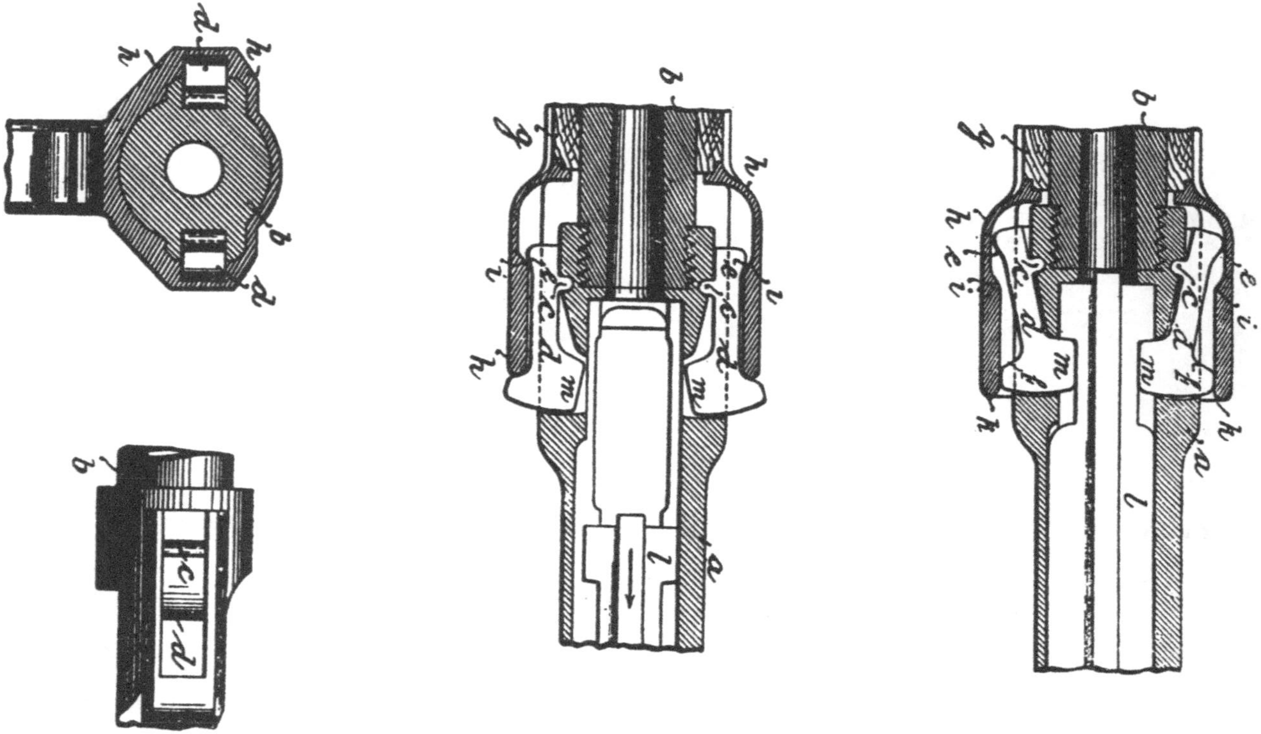

MAUSER "VALVE-TYPE" LOCK FOR SEMI-AUTOMATIC SHORT-RECOIL RIFLE

Top Line: Top view showing action locked. Rear ends of two locking arms are securely gripping cuts in breechblock. ("a" is receiver; "b" is barrel; "l" is breechblock; "m" and "m" are lock faces.)

Second Line: Top view showing action unlocked. As barrel and breechblock traveled back locked together, when the breech pressure dropped to safe limits the cam surfaces ("e" and "e") of the twin locks hit the slopes of the extension (at "i" and "i"), thereby camming the locks down the sides of the extension openings and out of engagement with the breechblock cuts. The barrel travel is halted at this point, but the impetus delivered to the breechblock is sufficient to carry that member back to eject, recock and reload.

Third Line: Front and side views of the locking elements.

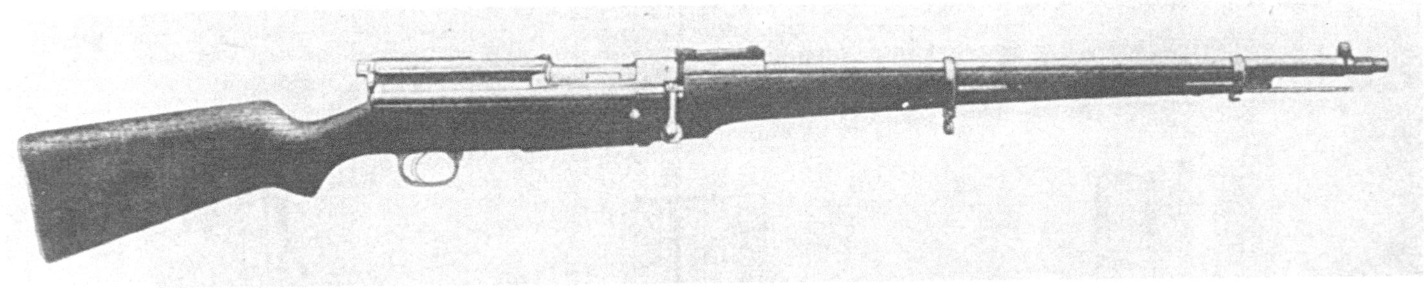

MODEL 1902, MAUSER SEMI-AUTOMATIC RIFLE, RECOIL OPERATED. RIGHT SIDE VIEW WITH ACTION CLOSED. This is a *long recoil* operated rifle with turning bolt lock. The design was not successful.

MODEL 1902, MAUSER SEMI-AUTOMATIC RIFLE. TOP VIEW WITH ACTION OPEN READY FOR LOADING AND COCKING HANDLE RETURNED TO FORWARD POSITION.

Note that this rifle is loaded with a clip through the top exactly as in the standard bolt action type of manually operated rifle.

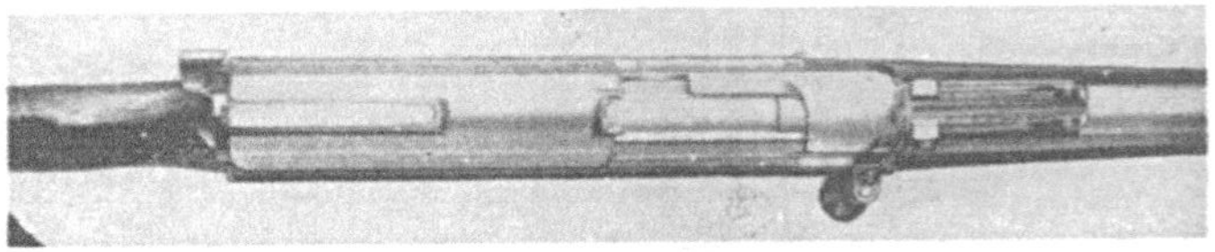

MODEL 1902, MAUSER SEMI-AUTOMATIC RIFLE. TOP VIEW WITH ACTION CLOSED AND COCKING HANDLE TURNED DOWN OUT OF ENGAGEMENT.

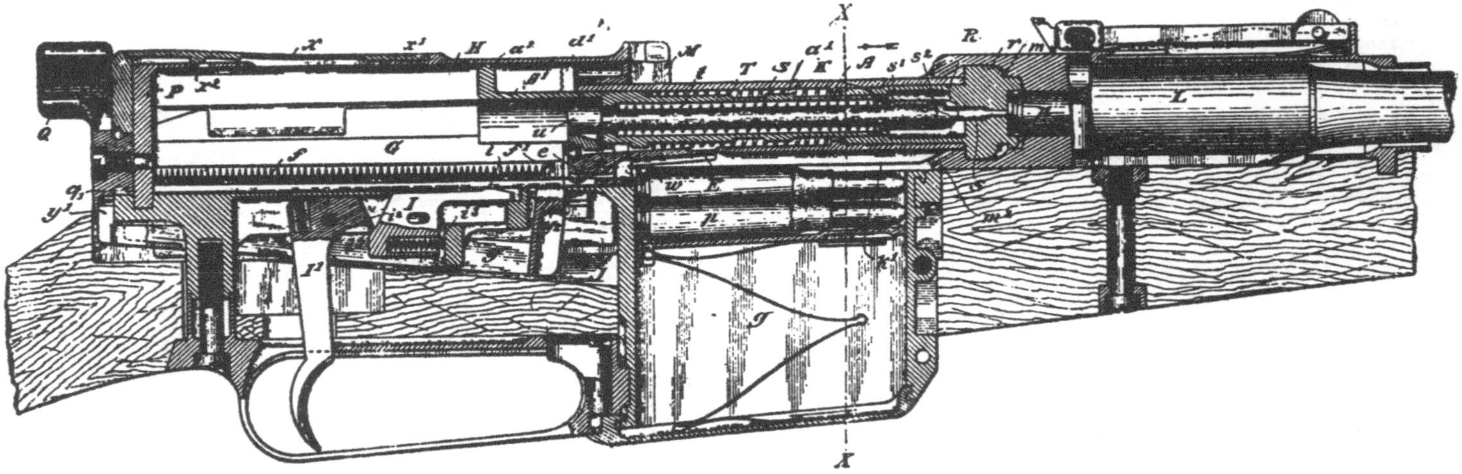

MAUSER SEMIAUTOMATIC RIFLE WITH LONG-RECOIL LOCKING SYSTEM

Right side view with magazine partly loaded. The cartridge in the chamber has just been hit by the striker.

All parts except the partly compressed magazine spring are shown at rest. Note the position of the locking lugs directly behind the head of the cartridge case.

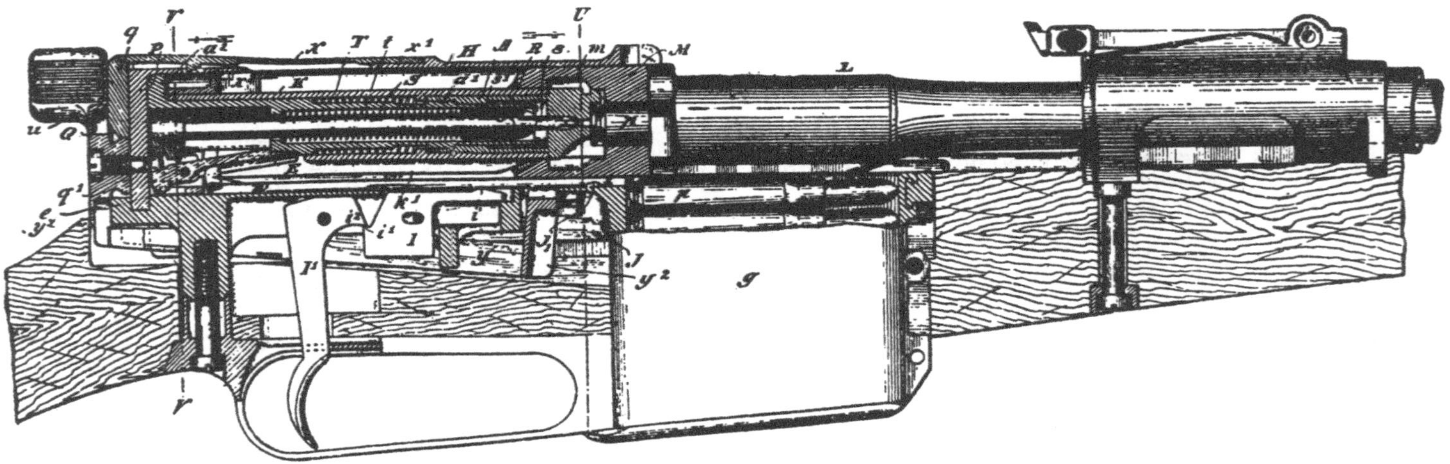

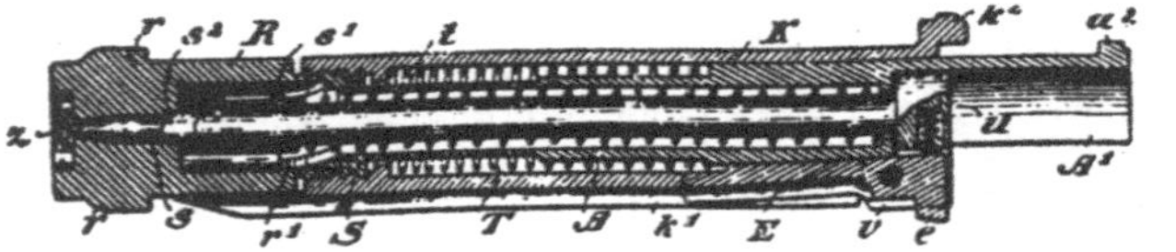

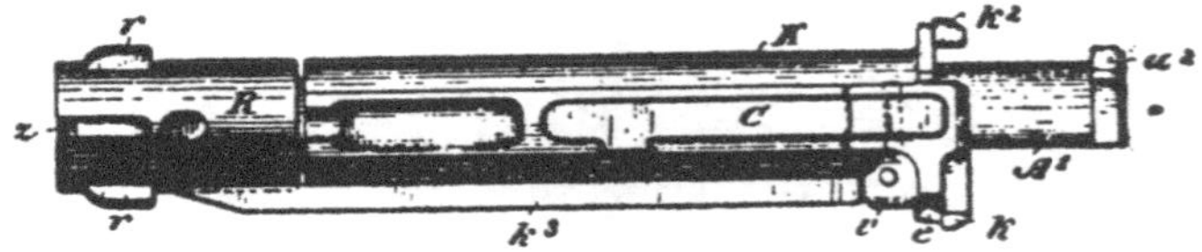

MAUSER SEMIAUTOMATIC RIFLE WITH LONG-RECOIL LOCKING SYSTEM

Above: Right side view showing barrel at rear end of its recoil stroke. In this design note that the barrel slides back *locked to the breech mechanism* until it passes completely over the magazine. At that point the breechblock is automatically held back by its catch, and the compressed barrel return spring pulls the barrel forward. While the rear section of the breechblock is held securely by the catch, the front bolt-head section is free to turn. The barrel pulling on the bolt head acts through the feather on the bolt to twist it until the lugs come free from their seats in the receiver, thereby unlocking the action so the barrel can complete its forward movement alone.

Below: Solid and sectional views of the breechblock showing its two-piece construction.

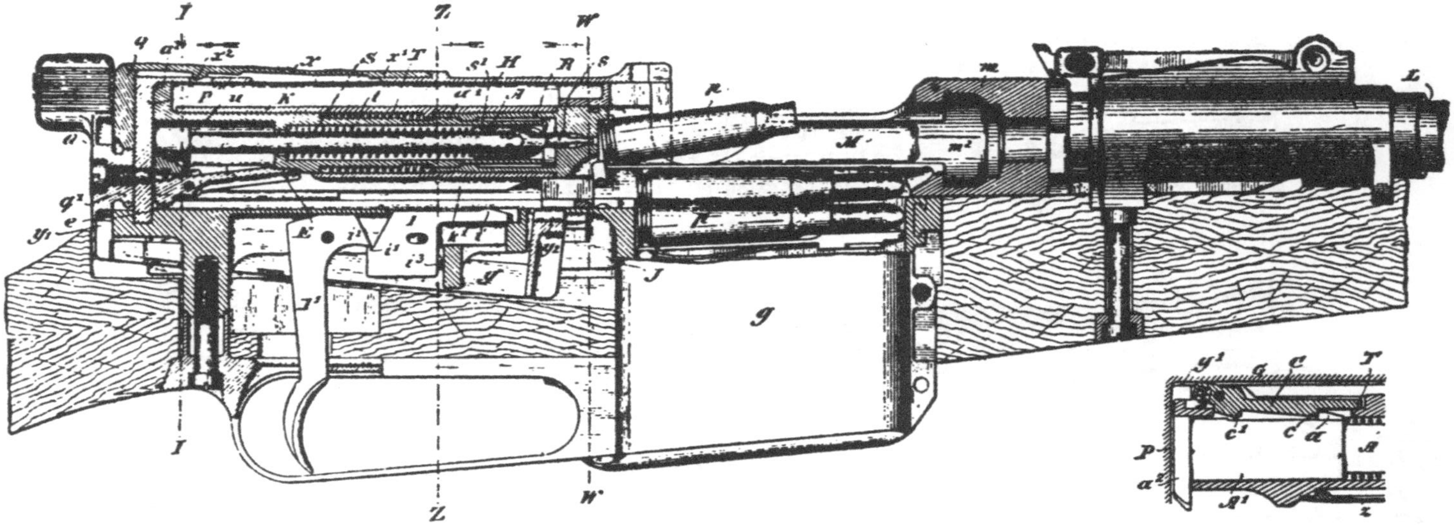

MAUSER SEMIAUTOMATIC RIFLE WITH LONG-RECOIL LOCKING SYSTEM

Right side view at moment of ejection. This is the third phase in long-recoil operation. The barrel has been returned by its spring to battery position ready to load. As the barrel was pulled forward from over the empty cartridge case, the ejector was actuated to hit the case held in the face of the breechblock by the extractor, resulting in ejection as shown. As the case leaves the rifle the breechblock catch is automatically freed. The compressed operating or recoil spring in the receiver behind the breechblock now starts to push that member forward. The sear catching in its bent in the striker, holds it in readiness to be cocked.

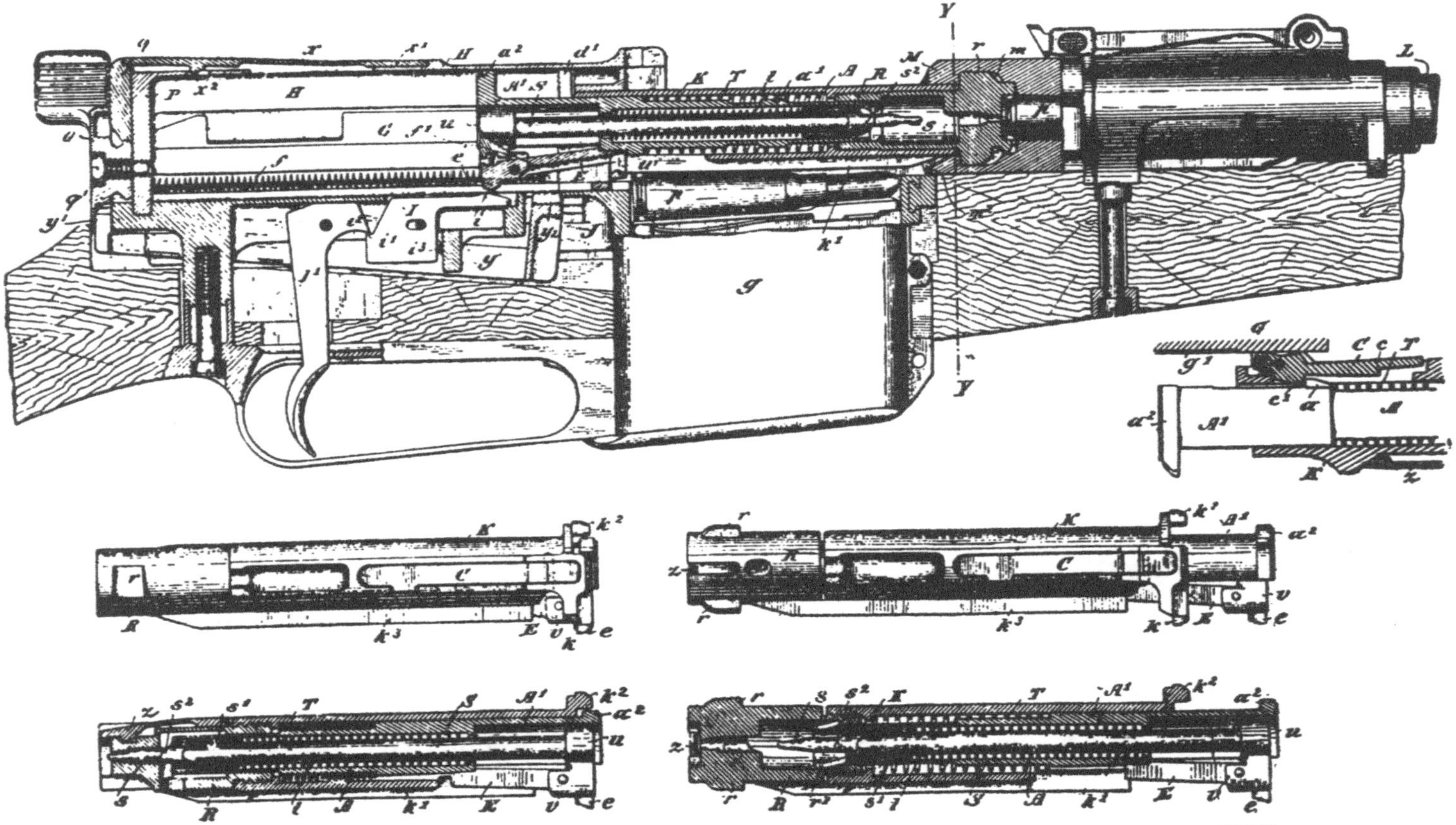

MAUSER SEMIAUTOMATIC RIFLE WITH LONG-RECOIL LOCKING SYSTEM

Top Line: Right side view showing arm with chamber reloaded ready for firing. When the breechblock moved forward, its face hit the top cartridge in the magazine and drove it to the firing chamber. The rear section of the breechblock exerting pressure on the rotating bolt head in its front end, caused the feather traveling in the receiver cam groove to revolve the bolt head into locking position at the close of the forward stroke. Pressure on the trigger will fire the cartridge in the chamber and start the automatic cycle of recoiling, unlocking, ejecting, reloading, relocking and cocking.

Second and Third Lines: Solid and sectional views in fired and unlocked position and also in cocked and locked position.

recoil spring drove the breechblock assembly forward. As the breechblock completed its forward movement, the projecting surfaces of the bolt locks hit surfaces on the barrel extension which cammed them in to engage in cuts in the side of the breechblock.

A standard type rotating hammer which struck the firing pin in the breechblock to discharge the arm constituted the firing mechanism in this design.

A special retaining catch was incorporated in some models to permit using the arm as a single shot rifle while holding the magazine in reserve. Other features were a special barrel stop or buffer, a simple firing pin safety working through a transversely displaceable stop, and a special extractor having separate extracting claw and spring.

Model 02 Mauser Semi-Automatic Rifle

This is a long-recoil operated rifle using military cartridges, but is of very complicated design.

A long-recoil automatic weapon may be generally defined as one in which the barrel and the breechblock recoil locked together for a greater distance than the overall length of the cartridge employed.

The barrel stays locked to the breech mechanism *for the full length of the unlocking stroke*. The breech mechanism is then caught and held automatically by a special catch while a barrel-return spring starts moving the barrel forward to closed position. The extraction, ejection and feeding of the new cartridge takes place *during forward movement of the barrel;* which requires that the only time the action is really open is when the barrel is in forward motion, or is fully home.

While this system has been employed successfully on automatic shot guns such as the Remington and the Winchester, it has operated successfully in production of rifles only in the Remington self-loading rifle invented by Browning, and the Belgian Browning.

The action is much more complicated and not nearly as satisfactory for high power rifle ammunition as the short-recoil type in which barrel travel is halted after short travel, and the action opens to extract and eject on the *rearward movement* and to load on the forward movement of the breechblock.

This Mauser Model 1902 has a true barrel which moves back and forth in an outer barrel or barrel casing. It is locked by two symmetrically placed lugs in the breechblock which are in vertical position when locked in their recesses in the rear of the barrel chamber.

The locking lugs are carried on the front of the breechblock. As the rifle recoils when fired, the barrel and breech mechanism recoil locked together compressing the recoil spring within the receiver.

At the end of the rearward stroke, the breechblock is caught and held by a catch, while the barrel is thrust forward by its return spring. This *pulling* motion of the barrel drawing with it the revolvable head of the breechblock, causes the head to turn until its lugs unlock from their seatings in the barrel behind the chamber. Thus the barrel can start forward over the top of the magazine with the empty cartridge case held firmly in the grip of the extractor in the bolt head. As the case clears the forward moving barrel, it is hit by the ejector and hurled out of the rifle.

The holding catch is then automatically released, permitting the breechblock to run forward and chamber a cartridge, meanwhile producing through feathers a helical movement. This turning movement of the bolt head causes it to engage it in its locking recesses behind the base of the cartridge. The firing mechanism is of the familiar striker type.

Model 06-08 Semiautomatic Rifle

The Model 06-08 while externally the same as the other semiautos, was issued with three different lock designs, as shown in the following pictures and drawings.

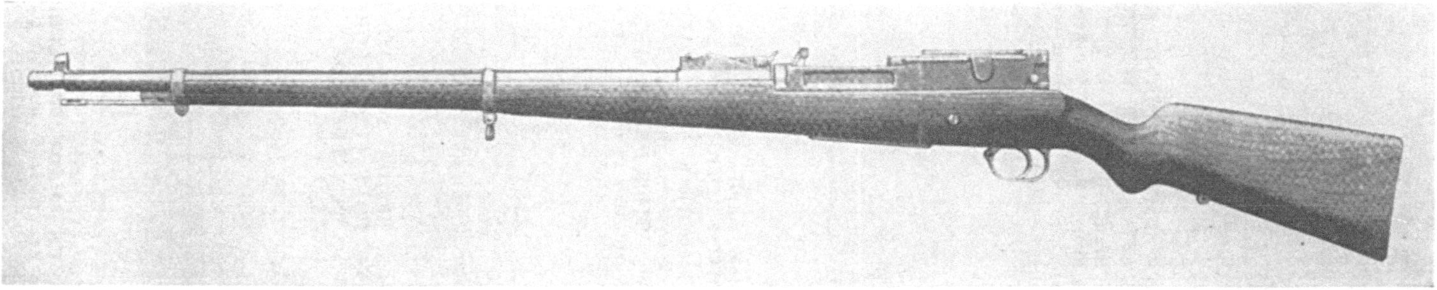

MODEL 06-08, MAUSER SEMI-AUTOMATIC RIFLE. LEFT SIDE VIEW WITH ACTION CLOSED READY TO FIRE.

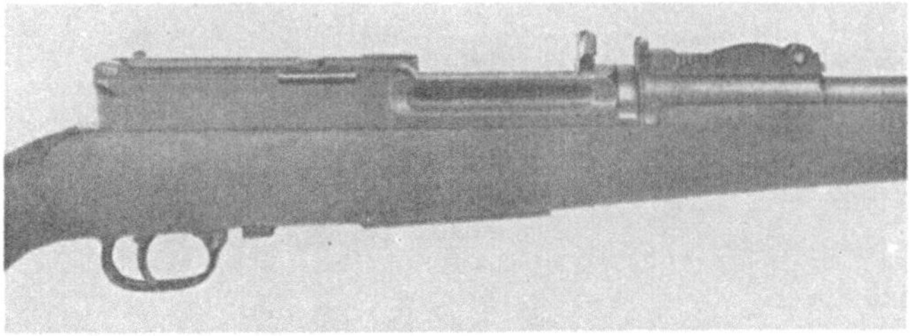

MODEL 06-08, MAUSER SEMI-AUTOMATIC RIFLE. RIGHT SIDE VIEW OF RECEIVER WITH ACTION CLOSED.

This model has the standard 5-shot magazine. Note that cocking handle on this arm is on top of the breechblock. The magazine release catch provided to permit insertion of an oversized magazine may be seen directly ahead of the triggerguard.

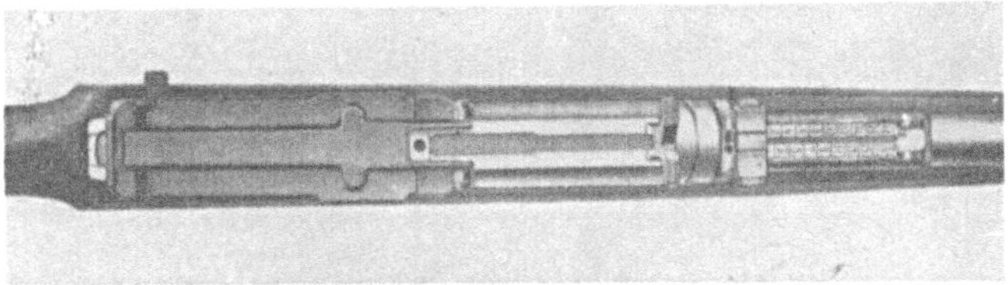

MODEL 06-08, MAUSER SEMI-AUTOMATIC RIFLE. TOP VIEW WITH ACTION CLOSED.

Top details of the breechblock which travels back in the receiver below the deck covering the rear section. The catch at the rear is the safety. Many features of this design were incorporated during World War II in German semi-automatic rifles.

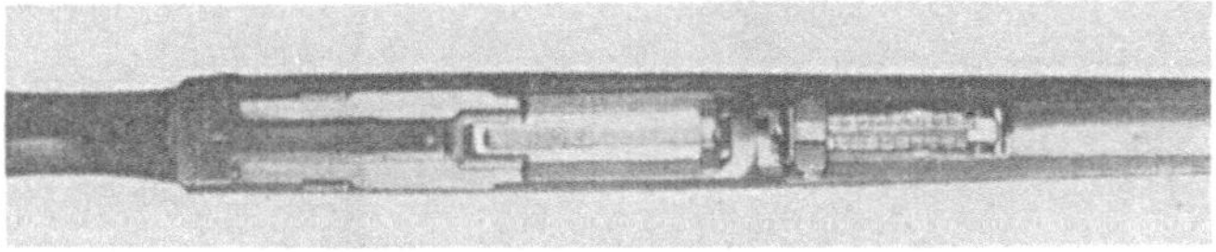

MODEL 06-08, MAUSER SEMI-AUTOMATIC RIFLE. TOP VIEW SHOWING BOLT FULLY FORWARD AND ACTION LOCKED.

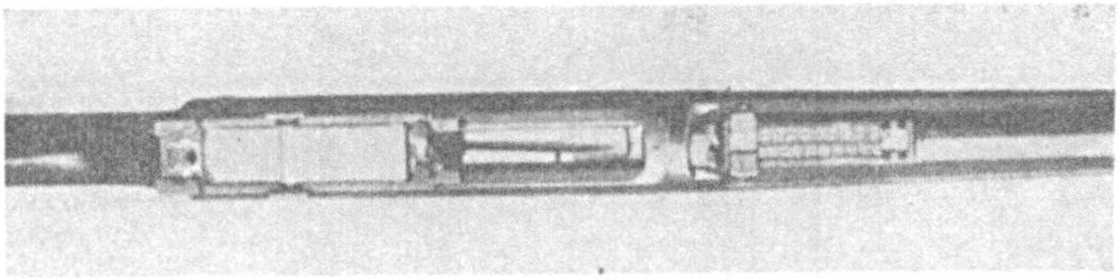

MODEL 06-08, MAUSER SEMI-AUTOMATIC RIFLE. TOP VIEW WITH ACTION OPEN SHOWING MAGAZINE FOLLOWER READY TO RECEIVE A CLIP OF LOADED CARTRIDGES.

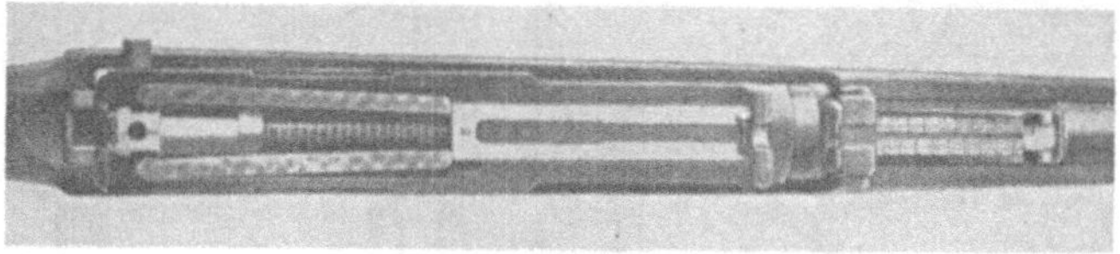

MODEL 06-08, MAUSER SEMI-AUTOMATIC RIFLE. TOP VIEW WITH DECK REMOVED AND ACTION CLOSED.

This photograph shows in detail the locking system of this arm. The bright surface bearing the figure "2" is the breechblock. The extractor is mounted in the top of the breechblock. A long metal locking arm is mounted to the rear of the breechblock on each side of the recoil spring and its guide. At the moment of firing these locking lugs hold the breechblock forward while they are pressed at their rear firmly against the side of the receiver wall by the heavy guide between them. In this system, the locking is too far back from the head of the cartridge case to be really satisfactory. This was purely an evolutionary design.

This was the "Selbstlade-mit Verriegelung durch Stützflappen." (Self-loader with lock through short flaps.) Barrel and breech mechanism recoil locked together. When pressure drops to safe limits, short steel flaps (or wings) below the breechblock are cammed out to push the locking levers out of the path of the breechblock. The barrel hits its stop and the breechblock travels back alone.

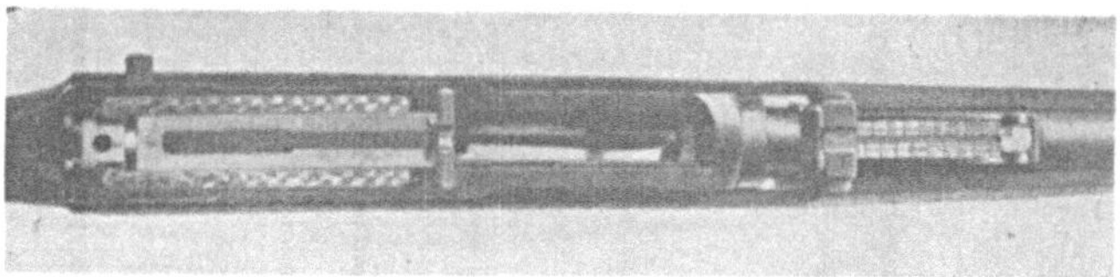

MODEL 06-08, MAUSER SEMI-AUTOMATIC RIFLE. TOP VIEW WITH DECK REMOVED AND FULL REAR POSITION.

A comparison of this photograph with that of the closed one will disclose the distance of barrel recoil by a comparison at the line of the rear sight.

At the instant of firing the barrel and breechblock are locked by the locking arms behind the breechblock being held firmly in place to support the breechblock-head. The unlocking flaps are hidden below the breechblock in this photo.

After short recoil during which pressure drops, the barrel hits its stop and is halted in rearward travel.

The forward ends of the locking arms are pushed outward, against the barrel extension by cam action of the short flaps as they turn and the breechblock continues to the rear to with the unlocked arms, indicated by the mottled steel tops, compress the recoil spring around its guide.

The photograph shows the breechblock in full rear position forced into their recesses in the barrel extension on either side of it.

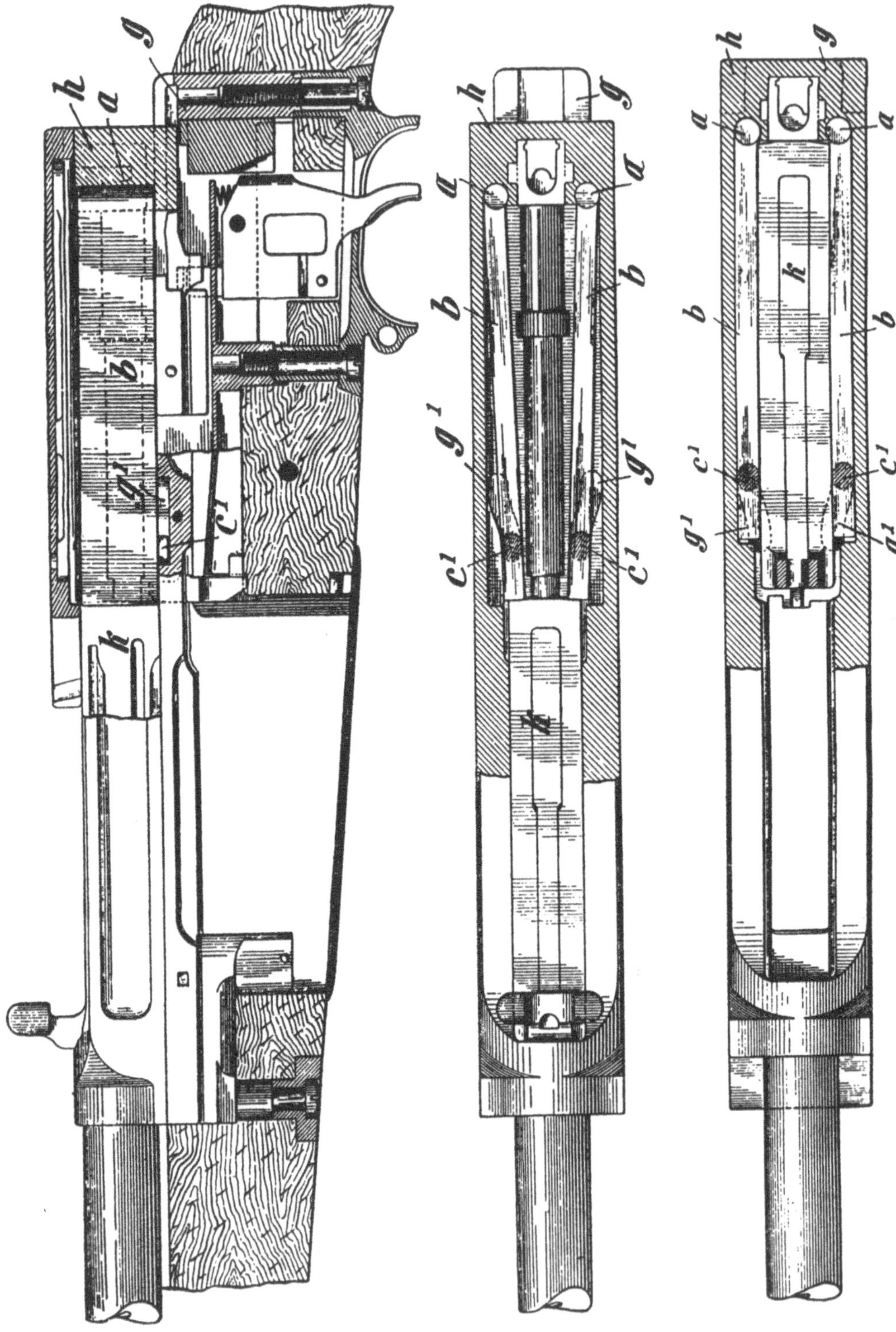
h
a
g
b
g^1
c^1
k

MAUSER 06-08 SEMI-AUTOMATIC RIFLE. DETAIL DRAWINGS OF THE LOCK.

(See page 152)

1. Side view of action closed showing details of trigger hook up.
2. Top phantom view showing action closed ready to fire. K is the breechblock which is locked. A and A are rear ends of the two bolt (or breechblock) locks, housed in recesses in the barrel extension on either side of the recoil spring guide. C and D are the center surfaces of the bolt locks G1 and G1 are the rear locking surfaces of the pivoting bolt lock releases. C1 and C1 are the forward faces of the short release (flap) locking cams in full forward position. H is a rear of the barrel extension and G is the receiver guide in which it travels.
3. Top phantom view showing the action open and the arm unlocked. Note that at the end of the recoil stroke, the two pieces G1-G1 have been swung completely around by the impinging faces of the cams and the reversal of position has forced the locking arms into their recesses in the barrel extension to permit the breechblock to release the barrel.

SPECIAL MAUSER SHORT-RECOIL LOCKING SYSTEM

(See page 154)

Top Line: Right side and rear detail of action. The action is fully forward and locked. Note that in this unusual design the lock is a heavy pivoting member which supports the actual breechblock at its rear. The guide tracks in the rear of the receiver indicate the distance of rearward travel of the action during recoil. ("c1" and "c2" are the lock cam faces which bring about the unlocking and locking movements).

Second Line: Right side view of weapon locked ready for firing, showing details of sear and striker construction.

Third Line: Right side views in full recoil position. (1) Side cutaway to show spring compression and ejector (f) details. (2) Details of lock member in unlocked position. On forward movement of breechblock, cam face "c2" will hit slope of receiver at "g2," thereby pivoting the lock down into engagement.

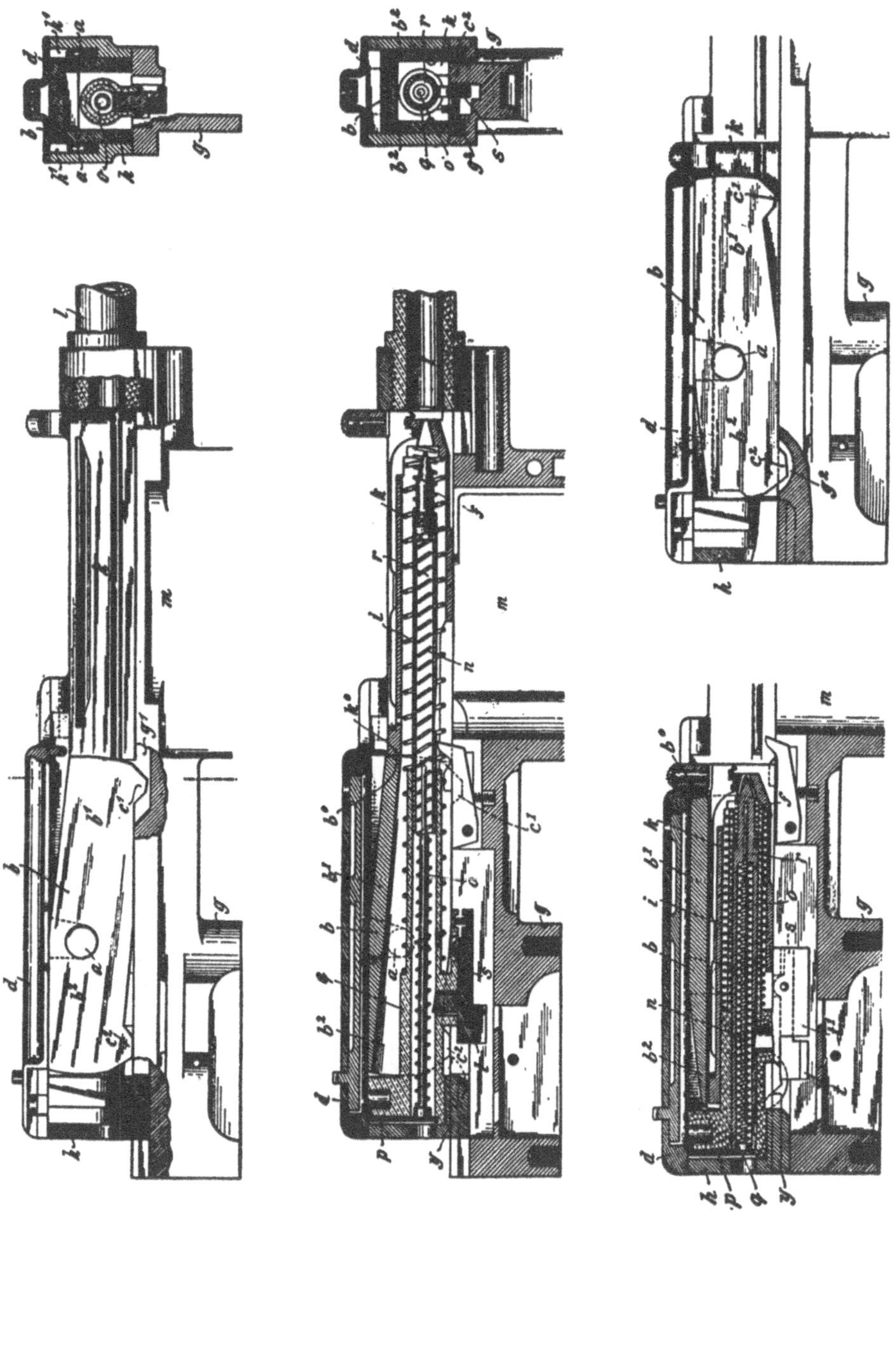

24. MAUSER SPORTING RIFLES

Genuine Mauser sporting rifles as made before and after World War I are among the finest weapons ever produced.

They were made in a very wide range of calibers, many of which are not manufactured in the United States.

The types and calibers generally marketed in this country are listed below:

Type A Sporter

This design was made in Germany specifically for foreign markets. It was finished much better than those for home consumption. Various types of sights were provided and this design was made in both single-trigger and double set-trigger types. The stock usually has a cheek-piece which is full checkered and a capped pistol grip. The fore-end tip is horn-capped. Sling swivels are provided, one on the barrel and the other on the stock. The usual barrel length of this type is 24-inches, round, although a later rifle with 22-inch barrel was also provided.

The standard calibers of the 24-inch was as follows: 7 x 57 mm (.276), 8 x 60 mm (.315), 9 x 57 mm (.355), 9.3 x 62 mm (.366), 10.75 x 68 mm (.423), .280 Ross, .30-06 U. S., .318 W.R., and .404 Eley.

With the 22-inch barrel calibers were: 6.5 x 54 mm (.256), and 8 x 51 mm (.315); also .250-3000 Savage.

SPORTING RIFLE

Mauser Cartridge Ballistics

Caliber	*Bullet Wt. gr.*	*Powder Wt. gr.*	*Cartridge Length inches*	*M.V.*	*M.E. f.p.*	*Traj. at 200 yards inches*
6.5x54 mm	119	33.2	2.68	2362	1468	4.4
8x51 mm	158	38.6	2.68	2380	1990	4.65
7x57 mm	154	50.1	3.07	2740	2567	2.52
8x60 mm	154	52.5	3.15	3000	3085	2.45
9x57 mm	247	46.3	3.03	2296	2885	4.60
9.3x62 mm	285	54.	3.92	2296	3486	4.40
10.75x68 mm	347	64.8	3.23	2313	4123	4.55
.250-3000 (Savage)	87	...	2.52	3080	1840	2.4
.280 Ross Rimless	150	50.	3.449	2870	2710	2.59
.30 U. S. '06	150	53.5	3.331	3000	2980	2.4
.318 W-R	250	52.	3.523	2230	2700	4.13
.404 Eley	400	82	3.504	2160	4173	3.9

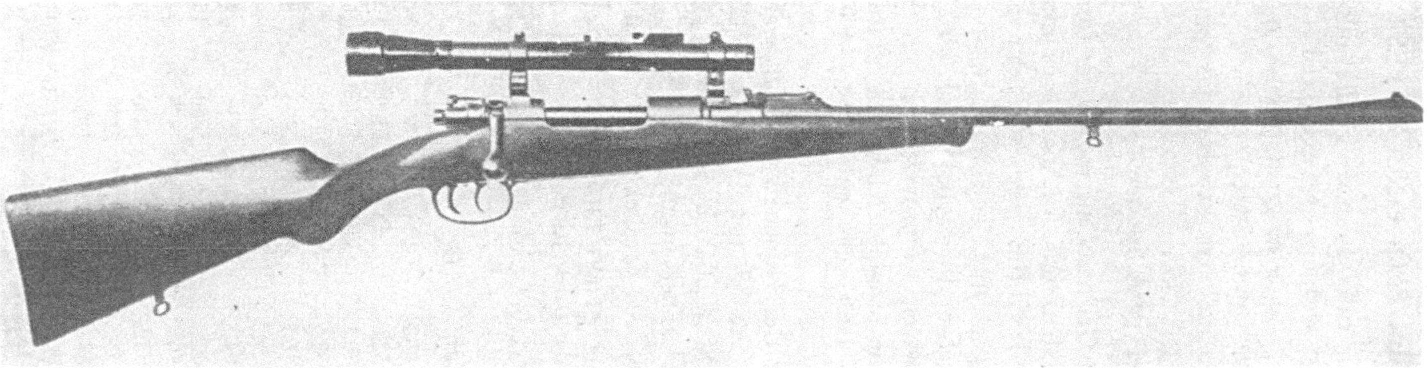

TYPE A MODEL 98-08. MAUSER MAGAZINE RIFLE WITH TELESCOPE

This Model with telescope and set triggers was designed for precision game shooting. Except for the triggers, it is mechanically the same as the Gewehr 98.

Stocks, calibers, general specifications and magazine floor plate release systems vary to a tremendous degree; but the mechanics are the same generally.

These rifles weighed about 7.25 to 7.75 pounds.

A table of ballistics for their cartridges is listed above. Ammunition when available may be obtained from the Stoeger Arms Corporation 507 Fifth Ave., New York, New York.

Type B Sporter

The Type B was manufactured in a variety of patterns.

They were made for the following cartridges: 7 x 57 mm (.276), 8 x 60 mm (.315), 9 x 57 mm (.355), 9.3 x 62 mm (.360), 10.75 x 68 mm (.423). The first pattern was supplied with 24-inch round barrel, silver bead foresight on matted block, pear-shaped bolt knob, hinged magazine bottom plate with lever release, walnut stock with cheek-piece, full checkered and steel capped pistol grip, hard rubber butt plate and sling swivels.

This type had dual set triggers. One model had leaf sights up to 300 yards, while another had a tangent curve sight up to 1000 yards.

The second pattern of this type B was the same as the first pattern except that a single trigger was used.

The third type was the same except that the barrel was half octagon. This was a set trigger model.

Pattern four had a single trigger and a full octagon barrel.

Pattern five was the same as pattern four with full octagon barrel but had set trigger.

Type K Sporter

The type K was a light weight short rifle with a 21.5-inch barrel. Other characteristics were the same as for the type A. It was made in single trigger and hair set-trigger type; also with round, partly round, partly octagon and full octagon barrels.

The type K was furnished only in calibers 6.5 x 54 mm (.256), 8 x 51 mm (.315), and .250-3000 Savage.

Note: The "K" action is identical with the standard but is shorter and lighter. "K" here means "kurz" or "short".

Type M Sporter

The type M was made with both single trigger and set triggers and with leaf sights or tangent curved sights. It is unusual in that the stock extends to the muzzle. The barrel is round and is 20-inches long. A silver bead from sight on a matted block is standard. The bolt handle is flat, another characteristic. The hinged magazine bottom plate has a lever release. The walnut stock has a checkered cheek

TYPE B MODEL 98-08. MAUSER MAGAZINE RIFLE. HUNTING TYPE WITH QUICK DETACHABLE MAGAZINE BOTTOM FOR RAPID MAGAZINE UNLOADING

In this type pushing a lever below the magazine bottom permits the bottom and magazine follower and spring to come out and permits removal of the cartridges without the necessity of working them through the action. Special hunting sights were furnished on this Model.

As modified after 1924, this design has the wide gas flange issued in that year.

Photograph from original Mauser records.

piece. The pistol grip is steel capped. The steel plate is ribbed and has a trap for cleaning rod which may be screwed together. Sling swivels are provided.

This model was made in the following calibers: 6.5 x 54 mm (.256), 7 x 57 mm (.276), 8 x 51 mm (.315), 8 x 60 mm (.315), and 9 x 57 mm (.355).

Type S Sporter

This model is practically the same as the type M except for checkering and bolt handle shape. This type was also made with flat bolt handle but may be encountered with pear shaped bolt handle. The wooden stock extends to the muzzle of the 20-inch barrel where it is joined with a steel cap. This type weighs about 7-pounds. For the U. S. trade it was manufactured in caliber .30-06 U. S. The selling price for this caliber was about $175, or about $25 higher than for the other calibers.

The Mauser Magnum Action

The so called "Magnum Action" is identical in all respects with the standard Mauser except for size and weight, being designed for super-power cartridges. It was used on type A rifles in calibers 10.75 x 68 mm and .404 Jeffery.

On special order Mauser-Magnum actions were also provided in the U. S. for the .280 Ross or the 318 W. R. Express. The American selling price of these Magnums was about $250.00. This is approximately double the cost of the standard action.

The only differences in design required by use of Magnum receivers develops where very large cartridges are used. When stacking in double or staggered rows would make the rifle too bulky, receivers are modified to use a box magazine extending below the line of the stock.

Receivers are all basically short, standard or Magnum lengths and weights, but each is specially milled for the cartridge used.

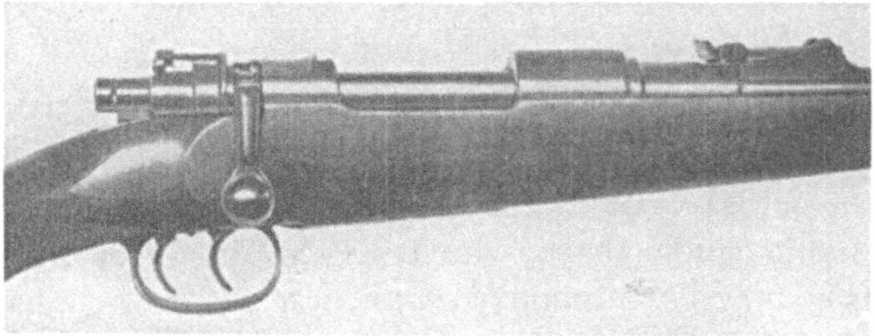

MODEL 98-08. MAUSER SPORTING RIFLE. RIGHT SIDE RECEIVER VIEW WITH ACTION CLOSED

This model is fitted with set triggers to permit an extremely light pull instead of the customary military two-stage trigger pull. The stock is of light weight sporting design and much better finished than the military issue.

MODEL 98-08. MAUSER HUNTING RIFLE WITH SET TRIGGERS. RIGHT SIDE VIEW WITH ACTION OPEN AND CLIP INSERTED IN GUIDE READY FOR MAGAZINE LOADING

When the cartridges have been stripped down by the thumb into the magazine, a forward thrust on the bolt handle will knock the empty clip out of the clip guide.

Note details of the famous third rear safety lug on bolt cylinder ahead of handle, and also of bolt travel guide rib on bolt cylinder.

Except for the set triggers, all mechanical parts of this model interchange with the standard German military type.

Note: Magnum actions and special sporting cartridge designs are not commonly clip loaded.

TYPE K MAUSER HUNTING RIFLE WITH SINGLE TRIGGER AND HUNTING SIGHTS

This type was developed with lever operated magazine bottom release, special forestock, sights and sling swivels for general hunting use.

"Kurz" or short actions are essentially reduced size military receivers and bolts. The original idea of using a shorter than standard action to reduce weight and speed operation was developed at Steyr in Austria, where 1910 Mexican and 1912 Serbian rifles were built.

The current Yugoslav military action is shorter than the standard Kar. 98k.

TYPE M MODEL 98-08. MAUSER HUNTING RIFLE

This Model varies from the previous ones only in furniture and sights. Note that it is equipped with sling swivels for carrying-sling and is stocked to the muzzle. This is an accepted feature in many fine European hunting rifles.

Recent 615 Models are the same as this except for wider gas flange. Photos are from original Mauser records.

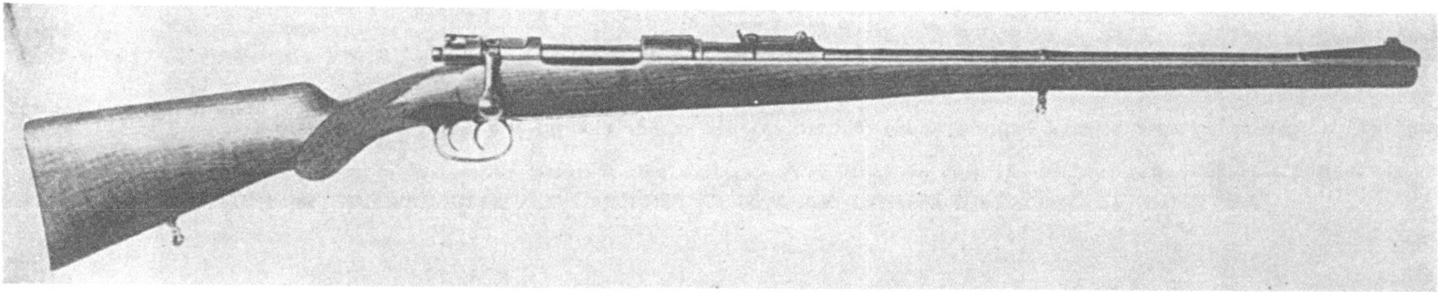

TYPE S MODEL 98-08. MAUSER SPORTING RIFLE, RIGHT SIDE VIEW

This is the original Mauser Sporting Rifle using the Model 98 action, mechanical parts of which are interchangeable with the German military types; but it is fitted with sporting sights and stocks and with double set triggers.

Though seldom seen in the U. S., this is a popular European sporting style. Calibers cover the entire range of foreign ammunition, since many of these weapons were barreled to order.

MODEL 98-08. MAUSER REPEATING RIFLE. ARMY TYPE TARGET RIFLE WITH LONG BARREL

This design was originally issued for target shooting and sniping. It is provided with the hunting-type magazine bottom release lever.

This type of rifle was designed for hunters seeking a sporter approximating their familiar service rifle in length, weight and general external design.

25. MAUSER .22 CALIBER SINGLE SHOT RIFLES

Mauser manufactured a large line of single shot and repeating bolt action .22 caliber rifles. Following is a list of the types:

1. Bolt action single shot with 20-inch round barrel and standing rear sight. No safety catch. Overall length 27 inches. Weight 4 pounds. This is a low priced rifle for sale at competitive prices. It is not intended for accurate target work.

2. The second model single shot has a 25.5-inch barrel with tangent curve sight graduated from 30 to 200 yards and wind gauge. The foresight is dovetailed into a raised block. This model is provided with sling swivels. It has a safety catch which is missing in the cheaper model. Overall length is 44-inches and weight about 6.5 pounds.

3. The single shot match rifle. This model has a 27.5-inch barrel, detachable screw rear sight with adjustable notch, safety catch, checkered pistol grip with horn cap, horn steel plate, and sling swivels. The overall length is 45.75 inches and the weight is 7.75 pounds. This type maye be fitted with a telescope sight. This is a precision rifle for target shooting.

Mauser .22 caliber arms are referred to in Europe as "Mauserlein", meaning diminutive, or small, Mausers.

26. MAUSER .22 CALIBER REPEATING RIFLES

Like the single shot rifle above, the .22 repeaters are all bolt action. As in the original military Mausers, a cam action of the bolt handle when opening the breech provides primary extraction to loosen the cartridge case, thereby assuring proper ejection.

The striker and firing pin, as in the case of the large Mauser, are in one piece, insuring the shortest firing stroke and time. The pull-off is the double military type and is adjustable.

In early models the rearward travel of the bolt was halted by the sear, but on improved models, a special bolt catch connected with the receiver was provided.

All genuine Mauser repeating .22 rifles have detachable box magazines, inserted in the receiver from below ahead of the triggerguard. They customarily hold 5-cartridges. These magazines have slots to show the number of cartridges remaining therein.

The M 410

This is a light model. The barrel is 24-inches long. A spring type open rear sight with screw adjustment for elevation and a bead foresight are standard. Sling swivels are provided, and a pistol grip which is full checkered. This arm has a safety catch. The overall length is 40.5 inches, and the weight about 4.5 pounds.

A heavy model of the above was made, fitted with tangent curve sight graduated from 30 to 200 yards, wind gauge, and having the bead front sight dovetailed into a raised block. The length is 43.5 inches, the weight about 6.75 pounds. This rifle sold in the United States for about $70.00.

The MS 350

This is a target repeater. This has a 26.5 inch round barrel. A 10-shot magazine was available. The action and barrel are grooved for sight and telescope. Sight may be adjustable micrometer or tangent. The foresights are driven in from the front and are interchangeable. A triangular, a bead and a blade front sight are supplied. A wing safety is provided on the bolt. The trigger and trigger guard are roughened. The walnut stock has a vulcanite steel plate and pistol grip. Vulcanite cap on pistol grip.

This is a standard European target rifle.

27. MAUSER-TYPE RIFLES

Rifles manufactured on the Mauser system were very common in Europe. In military calibers, they were manufactured by Fabrique Nationale at Herstal, in Belgium, particularly in calibers 7.9 mm—the standard German army cartridge. However, they were manufactured to order in all military calibers.

Weapons of this caliber and of design practically identical with the German Kar. 98 were also manufactured at the official arms works at Brno, in Czechoslovakia. These Czech rifles are of very superior manufacture, and were made in sporting as well as military type. Yugoslavia also built them from 1924 on.

Rifles of this same design for military purposes, were also manu factured at Radom, an official arsenal in Poland. The Polish weapons were excellent rifles, but crude in workmanship by comparison with high German standards.

Mauser rifles are also manufactured at the Spanish arsenal at Oviedo, Spain, in Mexico City, Mexico, and in the Argentine. Some were also manufactured in China. All these rifles were manufactured with equipment purchased from Germany, and in accordance with German specifications.

The Mexican Mauser in particular, is an excellent weapon made of the finest material and workmanship.

As already stated, the great Steyr Works, in Austria, also manufactured Mauser-type rifles in large quantity for export to foreign government.

The Schueler Mausers

In the field of large caliber high priced sporting rifles, the Mauser action has been adapted very widely. Perhaps the finest of these adaptations is that by Schueler at Suhl, Germany. Most of these rifles, like the genuine Waffenfabrik Mausers, have magazines containing 5-cartridges. Barrels may be partly octagon and partly round.

1. Type I has a 24-inch barrel with matted top rib. A tangent curve sight can be regulated up to 1000 meters. Other types of rear sights are used however. These rifles were manufactured for the following calibers; 6.5 x 54 mm (.256), 7 x 57 mm (.276), 8 x 57 mm (.315) and 9 x 57 mm (.355).

Other calibers in which some models of this rifle were manufactured are: 9.3 x 62 mm (.360) and 10.75 x 68 mm (.423).

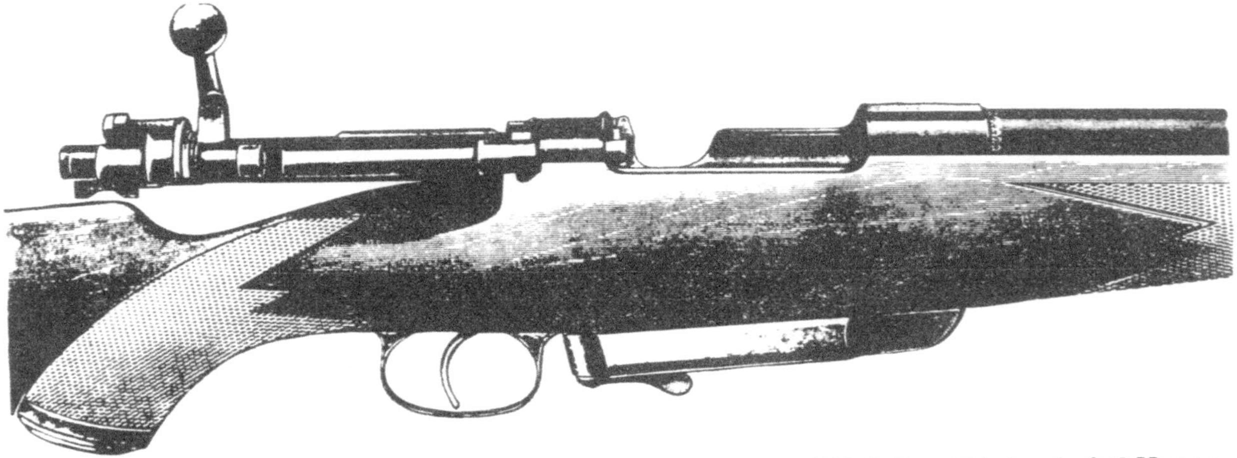

Schueler Magnum 500. Muzzle velocity 2460 fs. Striking energy 7230 ft. lbs. This is a typical Mauser-modified magnum action.

Projecting hinged box magazines are used where very large calibers make a standard box magazine impracticable. Generally, however, in design, Short, Standard and Magnum actions differ only in size. "Square" receiver bridges do not affect design. Receiver sizes are varied as required by cartridge dimensions.

2. Another rifle of this same design but of simpler finish with a round barrel was also made for the above cartridges. Special rifles in these types were chambered for .250-3000 Savage, .280 Ross, .30-06 U. S., .318 W. R., .404 British Eley, and for special German high power cartridges such as the 8 x 60 mm (.315) Magnum, and 9.3 x 62 mm (.366) Magnum.

3. Special "Schueler Magnum" rifles on the Mauser action were made in several types. One type with set trigger had a stock extending to the muzzle, while another had the customary sporting half-length fore-stock.

These rifles were manufactured for the following cartridges: 7 x 64 mm (.275) and 8 x 64 mm (.315) (7 x 64 is a special Magnum).

The "Schueler Magnum" magazine held only four cartridges, but a fifth could be inserted directly in the chamber. These rifles were made for the 11.2 x 72 mm (.441) Schueler cartridge, usually with set triggers. A modification of this rifle for elephant and rhino hunting, still more heavily built than the other, used the .500 Schueler cartridge. This cartridge had a muzzle velocity of 2460 feet per second with a striking energy of 7230 foot pounds. It is one of the most powerful rifles ever designed. The magazine extends below the bottom of the stock and has a lever release. This rifle was made with single trigger, and was designed for African and Indian big game shooting.

Schilling, Haenel and other manufacturers at Suhl made fair to good imitations of the genuine Mauser. Practically all of these were for standard German cartridges. Simson built fine rifles.

In England, Rigby made rifles on Mauser actions in calibers .275 High Velocity, .350 Magnum, .416 Big Game. These rifles retailed in the United States from $250 to about $400.00.

The British Jeffery rifles are also of Mauser pattern and use a series of special British cartridges including the .333 Jeffery, .375 Magnum, .400 Jeffery, .404 Jeffery, .450 No. 2, .475 No. 2, .500 Rimless, 577-3 inch, and .600.

These Jeffery's, are rifles of first line workmanship and quality and sold in the United States from $350.00 to $400.00 or more.

Other fine British manufacturers of rifles on the Mauser system, are Westley Richards (whose most famous model is the .425 W.R.) and Holland & Holland. H. & H. are particularly known for their .375

cartridge, and also for their .24 Apex. These rifles sell in the class of $275.00 and up. Vickers rifles are commonly .242 or .318.

Other Mauser Rifles

During World War I, the Germans manufactured a drill rifle having a cast receiver and a pipe barrel. There were *no working parts* in this model, which was intended strictly for instruction purposes.

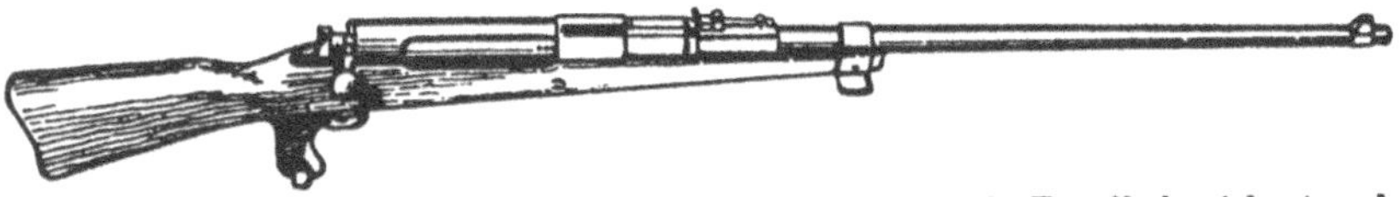

T-Gewehr als Einzellader (oben) und als Mehrlader mit gefederter Kolbenkappe (unten)

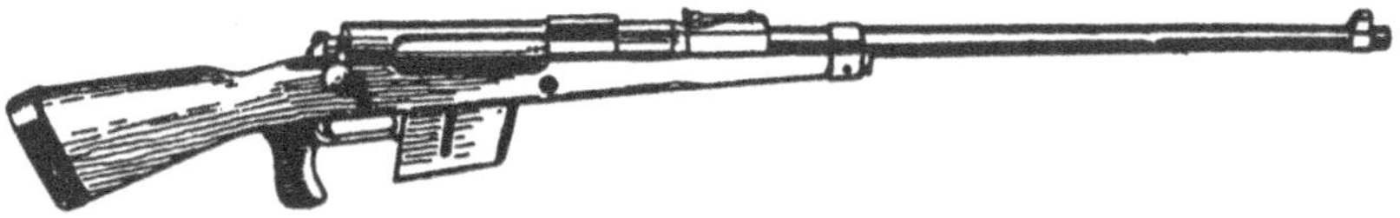

The German antitank rifle model 1918. Photo from the Mauser records

The German Anti Tank Rifle, Model 1918, was made at Oberndorf, in caliber 13 mm. It weighed 30 pounds, was 5.5-feet long, and was fired from a bipod. In its day it was a deadly and efficient weapon, capable of long range precision shooting. The cartridge it employed closely resembled the later developed famous United States .50 caliber Browning cartridge used in World War II.

In World War II Mauser designed and manufactured a limited number of semiauto gas-operated rifles known as 41-M. Gas expanding in a muzzle cone was trapped and some diverted to drive back a piston to unlock and function the action. This rifle was clip loaded. Capacity was 10 cartridges, 7.92 mm. The rifle was heavy and clumsy and the gas action fouled so badly that the design was soon abandoned.

28. MAUSER PISTOLS

Paul Mauser invented the first really successful military automatic pistol. While numerous experimental models had been manufactured at an earlier date, it was not until Mauser in 1896 patented his military automatic pistol that a weapon of reliable design was produced.

The diagram shown on page 172 is from the original specification filed by Mauser. It is remarkable in that this design was so right initially, that in all the years of manufacture from 1898 when it was first put on the market until 1945 when its manufacture was halted by the termination of World War II, only minor design changes were possible in this weapon.

In his patent specification Mauser described his invention and said that it had for its object, "A magazine repeating fire arm with a movable barrel, in which the recoil caused by the shock is used to unlock the bolt and open the breech, to eject the empty cartridge case and to work the firing mechanism, as well as to compress a number of springs arranged in such a manner as to effect the loading of a fresh cartridge, the reclosing of the breech and locking of the bolt and the advancing movement of the barrel. Upon these principles which are already partly known, I have devised a magazine fire arm in which all previous experiences in respect to this class of arm have been carefully taken into consideration, not only with regard to its ballistic qualities, but especially by the peculiar skillful construction of the component parts, and the manner of connecting them without the aid of screws; it may, therefore, be considered that the improved fire arm is in every way well adapted for military purposes."

This remarkable achievement of Mauser's in producing a pistol in which parts were made to interlock so that screws and pins were not necessary to its design has remained an outstanding achievement in firearms history.

The explanation given of the operation of the original drawing is as follows: "This is a vertical longitudinal section of an automatic repeating pistol. The barrel A has an extension A1 forming a breech casing for the breech bolt B, which is held forward by a spring supported by a pin A2 passing through a slot in the breech bolt. The breech bolt is locked in its closed position by a hinged locking piece C, having a pivot forming a projection against which the mainspring

D acts. When the pistol is discharged the barrel recoils a short distance, during which a projection on the locking piece C is drawn down into a recess in the block C1 by the action of the mainspring D, and the breech bolt is thereby unlocked. The barrel then stops, but the breech bolt B continues to recoil, cocking the hammer D1, opening the breech, extracting the spent cartridge, and allowing a fresh cartridge to rise from the magazine B1 into line with the breech bolt. The reaction of the breech bolt spring carries the bolt forward again to load the weapon and close the breech."

MODEL 96. MAUSER AUTOMATIC PISTOL WITH PROP-UP LOCKING BLOCK AND SHORT RECOILING BARREL

This is the original Mauser Military Model 7.63 mm caliber high velocity semi-automatic pistol.

The design of this arm was so fundamentally correct that after over 50 years of use, the latest model varies from the first model only in few details. The cartridge except for slightly improved ballistics is also the same.

This arm is a masterpiece of engineering design. It is very difficult to manufacture, but employs no holding pins. All parts fit into or are dovetailed into other parts. The only screw employed is the stock screw.

This was the first truly successful *military* model automatic pistol.

Quite as remarkable as the development of this pistol was the development of the cartridge which accompanied it. This cartridge is substantially the same today. Correctly known as the "7.63mm Mauser" it is often called the .30 Mauser. The actual caliber is .3008-inch.

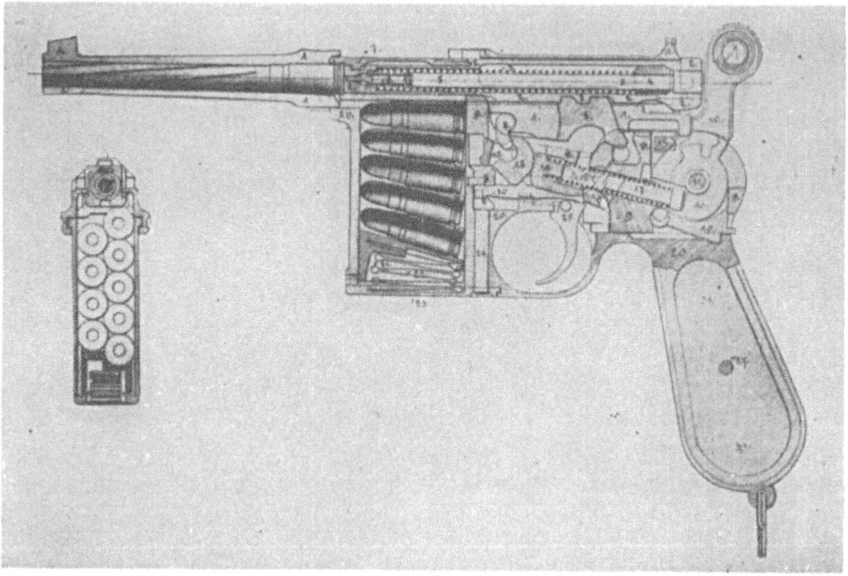

MODEL 96. MAUSER AUTOMATIC PISTOL. LEFT SIDE PHANTOM VIEW, AND REAR DETAIL OF MAGAZINE

This drawing shows all details of the weapon in closed position with all parts at rest except magazine spring which is under compression.

In the drawing an empty cartridge case is shown in the firing chamber. In actual operating conditions, this could not happen except at the instant of discharge. When this weapon is discharged, the barrel and breechblock recoil together and the empty cartridge case is ejected.

The barrel (A- at all points) and its rear barrel extension are forged from a single unit. The breechblock travels in the barrel extension. The front sight is also part of the barrel forging. Note that while there is a heavy recoil spring positioned around the firing pin, the firing pin itself has a small spring at its forward end within the heavier recoil spring which serves to withdraw the point of the firing pin from the primer of the fired cartridge as the weapon recoils. This prevents firing pin breakage. The locking block (8) is rising through the cut in the underside of the barrel extension to engage in its notches in the breechblock (2).

The detail view shows the construction of the magazine and its follower. As the cartridges are stripped down in the magazine through the top of the action, the shape of the follower forces the first cartridge to the right and zigzags each succeeding cartridge. Lips at the top of the magazine well prevent the cartridges from coming out accidentally.

To unload this weapon, it is necessary to move the breechblock back and forward once for each cartridge, which must be inserted in the chamber and ejected; or as an alternative method, the bottom of the magazine may be removed to permit withdrawal of the floor plate, magazine spring, follower and cartridges.

MODEL 96. MAUSER AUTOMATIC PISTOL. TABLE OF PARTS

A. This is the barrel, whose rear forging forms a barrel extension with guides for travel in the receiver. The rear of the extension is hollow to receive the breechblock inserted from the rear. In early models, such as this one, the rear sight was also part of the breechblock forging. In later models, separate rear sights were used.

2 The breechblock with gripping wings at its extreme rear protruding from its entrance to the barrel extension in which it is housed.

3 Recoil spring mounted around the striker.

4 Recoil spring compressor. This is a block inserted through the side of the barrel extension and through a slot in the breechblock. Its forward surface serves as a

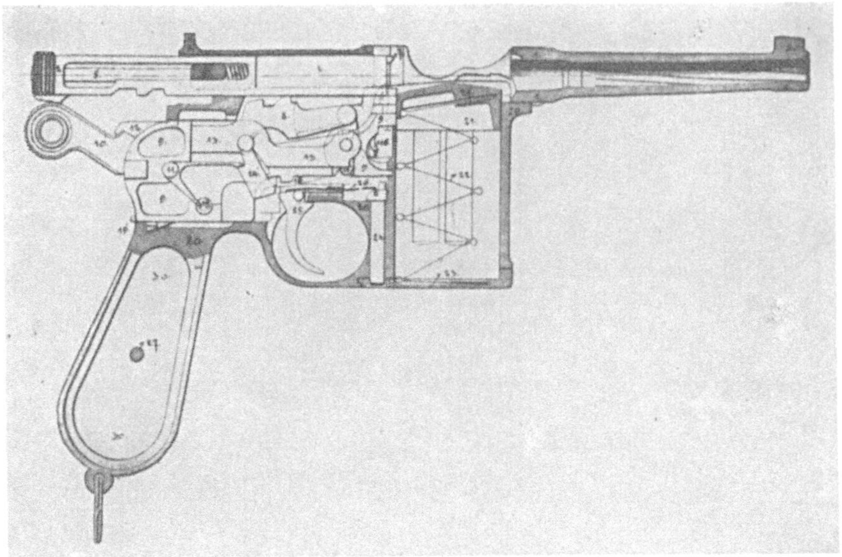

MODEL 96. MAUSER AUTOMATIC PISTOL. RIGHT SIDE VIEW WITH MAGAZINE EMPTY AND ACTION IN FULL OPEN POSITION READY FOR INSERTION OF A NEW LOADING CLIP IN THE FEED GUIDE

As a first step in loading this weapon, it is necessary to grasp the bolt wings at the rear of the breechblock and pull to the rear. The breechblock will ride over and cock the hammer, and the recoil spring will be compressed within the breechblock as shown in the drawing. The magazine follower will hold the breechblock in rear position. Inserting a clip in the clip guide and stripping the cartridges down into the magazine, then withdrawing the clip will release the breechblock and let it run forward to load the chamber ready for immediate firing.

---

compression point for the rear of the recoil spring when at rest. As the breechblock is drawn back, the recoil spring is compressed against this block.

5 **Firing** pin or striker.

6 **Firing** pin spring around point. Its diameter is reduced at its front end and it serves to draw the firing pin back into the face of the breechblock immediately after it has been driven forward by the blow of the hammer.

7 **Extractor.** This is a claw dovetailed into the top of the breechblock. It is of spring steel construction.

8 **Locking block.** It is slotted through its center and rests on a receiver projection directly below it. It is supported at its ends (at A and A) in projections on the sidewall of the receiver.

8-forward. The tooth marked "8" is a forward frontal projection of the lock. It is engaged by A8, the rocker below it.

A8 **Bolt rocker.** During the recoiling action, this rocker pressing on the tooth of the bolt lock produces the action which draws the locking piece of the piece 8 down out of engagement to unlock the weapon.

9 **Ejector.** This is a projection on the receiver in this arm.

9 lower and rear. These points indicate projections on the left side of the receiver in which parts are housed.

10 **Hammer.**

11 **Hammer axis pin.**

20 **Ribs at top edge of receiver in which barrel and breech mechanism slide.**

21 and 22 **Magazine follower and spring.**

23 **Magazine floor plate.**

24 **Magazine bottom attaching screw.**

20 **Parts of receiver.**

A7 **Mainspring and guide.**

21 **Stocks.** These are removable and are attached by the only screw used in the pistol.

Cartridge Specifications

The case is bottle-necked and rimless and positions on the shoulder in the chamber. The original muzzle velocity was about 1350 feet per second and with different loadings and different barrel lengths in modern ammunition it may run as high as 1475 feet per second. The breech pressure is extremely high for a pistol cartridge, being approximately 30,000 pounds per square inch. The cartridge has appeared with full patched bullets, soft nosed bullets and even hollow point bullets. Ammunition for this weapon was manufactured by all U. S. cartridge makers before World War II. The usual bullet weight is 85 grains with a cupra-nickel jacket.

MODEL 96. MAUSER AUTOMATIC PISTOL WITH SHOULDER STOCK ATTACHMENT

The rear of the pistol grip is slotted to permit an engagement lug at the forward end of the shoulder stock to be thrust up to act as a secure locking unit. The shoulder stock is usually of wood and is hollowed out to act as a holster for the pistol when not in use as a shoulder piece. The button seen at the rear of the stock is a spring catch which opens the hinged butt-plate to permit insertion of the pistol for carrying.

The range of this arm is in the neighborhood of 1,000 yards. Since the sighting arrangements for use as a pistol and use as a shoulder arm are quite different, best results cannot be obtained by attempting to use the weapon as a dual purpose arm. However, when sights are properly adjusted for shoulder firing, this is an effective arm at very much greater than ordinary pistol ranges.

The stocks shown are those *originally* used by Mauser and are *not* typical of the plain grooved wood stocks later supplied.

This pistol was very extensively used by the Boers in South Africa in 1898, shortly after its introduction, and was considered a very formidable weapon. The magazine holds 5 or 10 cartridges which can be stripped in through the top of the action from a clip (exactly as in the Mauser rifle) into the box magazine ahead of the trigger.

When the bolt is pulled back it cocks the hammer and the magazine follower rises to hold the action open for loading. A clip of cartridges is inserted in the clip guides and the cartridges pushed down into the magazine. When the clip is pulled up out of the top of the guides the compressed spring drives the bolt forward to chamber a cartridge, leaving the weapon ready to be fired. In original tests it was fired at the rate of 6 shots per second and better than 80 shots per minute. The extreme range is well over 1000 yards and the pistol was provided with a detachable shoulder stock which served as a holster for the pistol itself when not in use.

The original 1898 production had a 4.75 inch barrel including the length of the chamber. The hammer is large and rises into the line of sight as it hits the firing pin. This blocks the line of vision of the

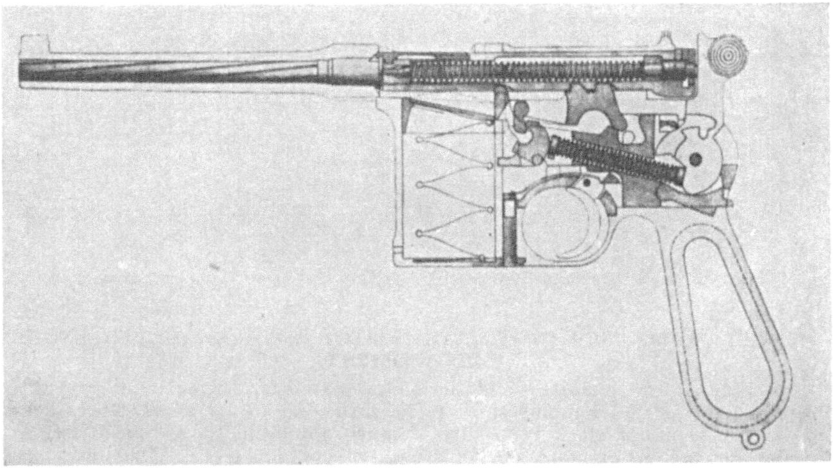

MODEL 96. MAUSER AUTOMATIC PISTOL. LEFT SIDE PHANTOM VIEW WITH ACTION CLOSED AND ALL PARTS AT REST

The barrel and barrel extension, which travels in the receiver and which houses the firing assembly, are forged from a single unit in the genuine Mauser manufacture. At the instant of recoil, the barrel and the breechblock are securely locked by action of the toothed prop-up locking block seen with its teeth engaging in the underside of the breechblock. These teeth rise through a cut in the underside of the barrel extension to engage with notches in the underside of the breechblock when the action is fully home.

During the opening recoil movement as the weapon is discharged, the barrel and extension travels back locked with the breechblock. When the barrel hits its stop and its travel is halted, the prop-up block is far enough back that it can be cammed down out of engagement. This leaves the breechblock free to travel to the rear, to extract, eject and ride over and cock the hammer. The heavy coil spring positioned around the striker is *not* a striker spring. It is the recoil spring which serves to drive the breechblock forward on the loading stroke, since it is compressed during rearward movement of the breechblock. The coil spring with guide below the hammer is the mainspring. A small striker return spring is positioned just to the rear of the striker point.

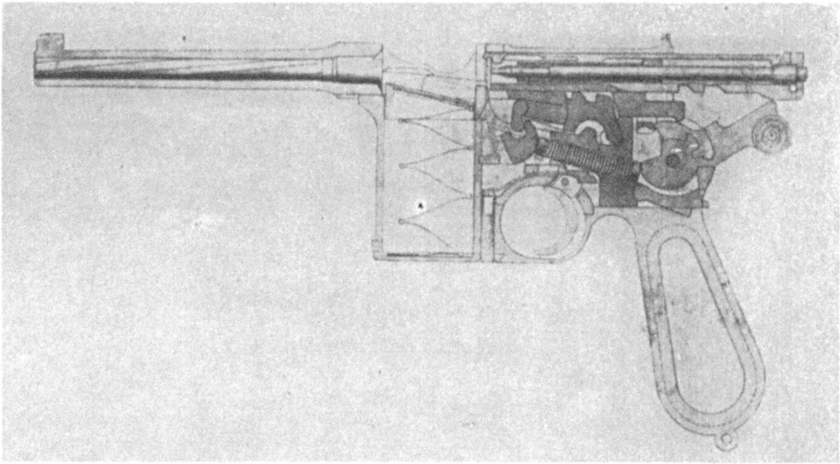

MODEL 96. MAUSER AUTOMATIC PISTOL. LEFT SIDE PHANTOM VIEW SHOWING ACTION OPEN AND MAGAZINE EMPTY

When the last cartridge has been fired, the magazine follower rises, and prevents forward movement of the retracted breechblock.

A comparison with the closed drawing will show the length of recoil of the barrel. The recoil spring is seen held compressed ready to drive the breechblock forward when released. The locking block has been cammed down, and the engagement notches in the underside of the breechblock may be seen.

The breechblock riding back over the hammer has rotated it about its axis pin causing its toe to thrust forward against the mainspring guide and thereby compress the mainspring.

When the weapon is open in this condition, a clip loaded with 10 cartridges is inserted in the feed guide. The cartridges are stripped down into the magazine with the thumb exactly as in the case of a Mauser rifle.

When the clip is pulled up out of the action, since the magazine follower is now down in the magazine well with cartridges on top of it, the breechblock is free and the recoil spring will drive it forward to chamber a cartridge.

On each successive pull of the trigger the hammer will be released to strike the firing pin and discharge the cartridge in the chamber. The barrel and breechblock will recoil until the barrel reaches its stop and is halted. At that point the locking block will be cammed down out of contact with its engagement notches in the underside of the breechblock.

The breechblock will then continue to the rear in its guide in the barrel extension carrying the empty cartridge case held in the extractor in its face, seen at the top of the front of the breechblock. The case hitting the ejector will be hurled out of the pistol. The compressed magazine spring will force the cartridges up to bring the next cartridge in line for loading, the cartridge being held in place by a lip at the top of the magazine well. The rearward stroke of the breechblock will cock the hammer. At the end of the recoil stroke, the recoil spring will reassert itself and drive the breechblock forward to chamber a cartridge and snap the extractor into the cannelure in the cartridge case for extraction. When the trigger is released, its spring will move it forward far enough to effect contact with sear to permit release of the hammer for firing the next shot. Note that this model is *semi-automatic* in that a pull is necessary on the trigger to fire each individual shot. This is a very desirable feature as it prevents firing more than one shot on a trigger pull, thereby preventing accidents. A full automatic pistol fired from one hand is not a practical weapon, since it cannot be held effectively on a target.

MODEL 96. MAUSER AUTOMATIC PISTOL WITH SHORT BARREL

This arm is identical with the one already described except for a short barrel. Arms of this type were originally designed for police work, while the longer ones were intended as military pistols. Magazine capacity was 10-shots in the pistol shown; but arms of similar design were also made with 5-shot box magazines.

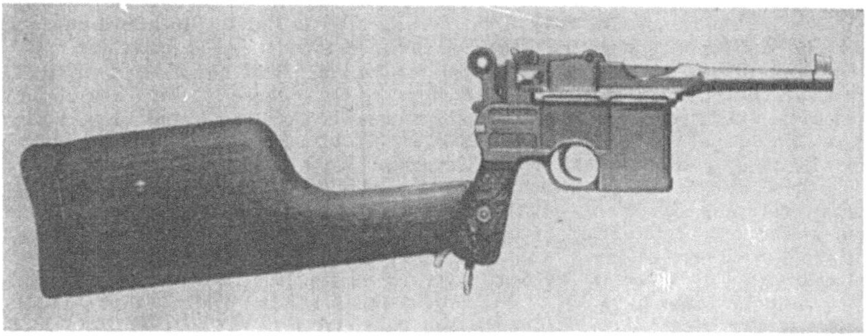

MODEL 96. MAUSER AUTOMATIC PISTOL WITH SHORT BARREL AND SHOULDER STOCK

When the short barrel model was issued, an improved form of shoulder stock with lightened wooden butt as shown was introduced. Except for minor details, however, it is identical with the original model.

rear sight except when the pistol is at full cock. The rear sight is a plain V which is not adjustable.

The front sight is part of the barrel forging itself. A thumb safety on the left side of the receiver is pushed down to lock the hammer safety when the weapon is cocked.

This original design was altered very shortly after its introduction, being replaced with a model having a 5.5 inch barrel, a lower hammer which does not interfere with the line of sight, and a tangent rear sight, graduated from 50 to 700 meters and mounted above the barrel extension. This was the weapon widely used in the Boer War.

One of the earliest and most enthusiastic users of the original Mauser pistol was Winston Churchill, who at that time was a Lieutenant in the 21st Lancers. He carried the pistol at the battle of Omdurman, and in his history of that campaign, gives a graphic account of how he was able to shoot his way out of a native trap because of the large magazine capacity and the rapidity of fire. His stiring account of the episode is one of the classic stories of the Sudan campaign. These early models use 4-groove rifling. In 1905 a 4-inch barrel model, having a somewhat squarer rib than usual, a 5-shot magazine, and a very large hammer, with an exceptionally large hammer hole was introduced in small quantity. This too, had 4-groove right twist rifling.

The 1912 and Subsequent Models

In 1912, a 10-shot model with a 5.5-inch barrel was reintroduced. The rifling was changed to 6 grooves, right hand twist, having very narrow lands. Some slight details in the action were changed at this time. A broader and shorter extractor was included, a smaller hammer, the safety lever was designed to be pushed *upwards* to produce the safe position, the striker (which heretofore had been retained by a sliding plate dovetailed into the rear of the bolt) now had two lugs turned on its head, the head itself being cross cut to accept the point of the cleaning rod furnished with the pistol—this was used to remove the striker when necessary. Pushing the striker in against this spring and giving a half turn to the right disengages the lugs, permitting the striker to be withdrawn.

Shortly afterwards other minor changes were made. The T section of the rear of the bolt were made broader, its under lugs which serve to rock the hammer back, was made larger and longer and extended to the rear locking recess, the front mainspring plunger was pierced to take the rocker (prior to this time the rocker had rested in front of the plunger). The front plunger was given a tail to project through the claw of the receiver as far as the sear lever to serve as a disconnector. While the front plunger is pressed back by the rocker, the tail prevents the sear lever from being elevated by the trigger and also serves to hold the front plunger.

The pivot of the safety lever engages directly in the safety bent on the hammer instead of using a separate bar as heretofore. When it is pushed upwards, a projection on the pivot enters the safety bent, a slot extending from the recess straight upwards. A second bent is provided to receive the pivot of the safety lever when the hammer is down. This feature serves to prevent the arm from being cocked if the safety is applied.

The trigger and its spring were mounted directly on the receiver (heretofore they had been on a separate block). This requires only a narrow slot in the claw of the receiver above the triggerguard.

This variety fitted with a tangent sight graduated from 50 to 1000 meters, was very extensively used during World War I. Although never an *official* German pistol, it was widely manufactured and issued to troops as a *substitute* weapon.

World War I Conversion

During World War I, large numbers of this model were converted to take the standard German army cartridge, for the Pistole 08, or 9 millimeter Luger cartridge as it is known in this country. These pistols are distinguishable from the standard 7.63mm type by the figure "9" usually painted or burned into the grips. This type of pistol used the standard 9mm Luger cartridge as manufactured in the United States, prior to World War II.

With the ending of World War I, the Mauser factory was not permitted under the terms of the treaty of Versailles to manufacture the 7.63mm with the standard barrel. By a subterfuge, it was decided that the pistol might be manufactured with a barrel a fraction under 4 inches, including the chamber length. This model had the same action as the standard Mauser used during the War, but its grip was modeled after the old 1905 model. This type will often be found with Swiss markings. It was also widely distributed in the United States as an export weapon, fitted with specially shaped Franzite grips. It has the standard 50 to 1000 meter sight. This model, introduced shortly after 1920, is often popularly referred to as the "Bolo" Mauser from the fact that very large numbers were shipped into Russia, where they were popular with the Bolshevik group striving for power at that time.

At this juncture it might be well to point out that this pistol in this caliber from its very inception was very widely accepted throughout all parts of Russia and Siberia. It was and is one of the most popular weapons ever used in the Orient. The official Russian army automatic pistol cartridge today is the 7.63mm Mauser cartridge with a somewhat lighter load, although officially called "7.62 mm."

The next alteration was in 1926 when the barrel length was set at 5.25 inches which kept it in the specifications set forth in the Versailles Treaty.

The changes made at that time include the following: The safety lock was altered to require pushing the thumbpiece up to set the safety, though it worked on the same principle as heretofore. The pivot pierces the side of the receiver and enters a recess in the side of the hammer to serve as a safety bar. This hammer recess is cut to enable the hammer to be released when the safety is applied but to halt the movement before it can touch the firing pin head. This is a so-called "universal" safety to permit the hammer to be lowered in complete safety on a loaded chamber.

The sear is extended further back on the left outside the receiver and enters a recess in the safety lever so cut as to permit sear movement only when at full fire or at full safe position. Utilizing this system, it is possible to load the firing chamber, set the safety and then merely press the trigger to drop the hammer safely. Pushing the safety off, leaves the arm ready to be cocked for firing the chamber.

The Mauser Universal Safety Lock was officially included in all models later manufactured by Mauser. This model was widely marketed as the Model 1930.

This universal safety lock will also be found on many of the short Bolo Mausers, indicating that production was continued in the 1930's.

29. 9MM MAUSER, MILITARY PISTOL

In 1908 Mauser introduced this pistol to shoot a *special* long 9mm cartridge. Because this model was used largely for shipment to Africa, South America and the Orient, where a larger caliber than the original 7.63mm was desired, it is frequently referred to as the "Export Mauser." Enough of them reached the United States to encourage our ammunition makers to produce the special 9mm cartridges required for this pistol; but manufacture has been discontinued for a number of years.

There is no difference in the two models except in the chamber, bore of the barrel, and sights. The description of the 7.63mm covers all features of the 9mm.

The cartridge case is straight-sided instead of bottle-necked but is the same length and head diameter as the 7.63mm. These cartridges were loaded on 10-shot clips exactly as in the case of the smaller caliber.

The bullet weighs 128 grains and may be cupro-nickel or steel jacketed. The original muzzle velocity was about 1300 to 1350 feet per second. This cartridge was never adapted to any other automatic pistol, although it was used in some European submachine guns.

This arm is a perfect example of Mauser's system of operation. First, he designed the pistol in caliber 7.63mm to meet German requirements; next he developed a *special* 9mm cartridge which could be used in a pistol made on the same machinery and to the same specifications (except for bore and sights) as the smaller caliber; hence when it became necessary to convert the Mauser for German military use to shoot the totally different 9mm Luger cartridge, it was possible to do this with little more than a change of bore and a minor magazine adjustment.

All Mauser's arms and ammunition were created on this general system, which allowed the Germans to fill foreign orders in times of peace to keep up their factory potential; and enabled them to swing into high gear for German military production on the shortest notice.

30. 9MM PARABELLUM, MAUSER MILITARY PISTOL

When a Mauser Military Pistol of the general 1898 design is encountered with the figure "9" painted on or burned into the stocks, it indicates a standard Mauser pistol with magazine and barrel adapted to handle the regular 9mm Luger cartridge known in Germany as the "cartridge for Pistole 08." In other parts of the world it is called "9mm Parabellum" (Latin "for war"). This powerful 9mm cartridge must not be confused with the common 9mm short used in pocket pistols.

During World War I, because of a shortage of Luger pistols and a desire to standardize ammunition insofar as was possible, the Germans utilized Mauser machinery to manufacture these pistols for their regular service ammunition.

Except for such differences as were necessitated by the increase in caliber, this model is identical with the standard 7.63mm type.

The tangent rear sight on this model is graduated only to 500 meters because of the lower muzzle velocity (about 1025 feet per second) of the Parabellum cartridge, as compared with the 7.63mm Mauser (approximately 1375 feet per second). The effective range of the Mauser cartridge when used with a shoulder stock weapon is of course, considerably greater than that of the Luger (or Parabellum) cartridge.

Since the barrel is the standard 7.63mm bored out to 9mm, its walls are not as thick in the larger caliber, though the *external* diameters are the same.

When these arms are encountered with the stamp of the Prussian Government on the barrel and on the receiver, they indicate weapons made between 1916 and the end of World War I as officers' service pistols. Arms of this caliber which do *not* bear the government stamp but *do* bear standard German proof marks, were assembled from spare parts or manufactured in contravention of the rules of the League of Nations as set up by the Treaty of Versailles after World War I. Large numbers of these arms were exported both to South America and to the Far East. Those found in the United States as a rule will be the official Prussian Goverment issue, which were brought here as War souvenirs.

31. MODEL 712, MAUSER FULL AUTOMATIC PISTOL, CALIBER 7.63MM MAUSER

This arm (known in Germany as the "Schnell-Feuer-Pistole") is a full and semi-automatic weapon which is not strictly a pistol. It is more directly classifiable as a submachine gun.

It should be emphasized that its unlicensed ownership in the U. S. is a criminal offense, punishable under Federal Law. It must be registered with the Federal Authorities.

This arm is a modification of the original Military Model 7.63mm Mauser, and partakes of most of its general characteristics.

It was manufactured with a 10-shot box type magazine flush with the magazine housing (the bottom of the triggerguard) and also with a 20 shot magazine. Both are detachable.

When loaded through the top with a 10-shot clip, it loads in practically standard Mauser pistol fashion. However, it may also be loaded by inserting a loaded magazine from below. The breechblock must then be pulled back and allowed to move forward to chamber a cartridge. (When this arm is loaded from a clip through the top of the open action, the breechblock moves forward to chamber a cartridge when the clip is removed *and* the breechblock released from its rear open position.)

When loading this model, pulling the breechblock back cocks the hammer, which then rises into a cut in the underside of the breechblock to hold it open. Drawing back the hammer slightly with the thumb will ease it out of its notch in the underside of the breechblock to release that member. (Note that this differs from the original pistol in which removal of the clip alone releases the breechblock).

When the last cartridge has been fired, the breechblock is held back by a projecting stud on the carrier which leaves the pistol ready to receive a new magazine and also warns that the arm is empty.

In loading these later models, when the action is open, two successive clips of 10 cartridges each may be stripped down from above through the open action.

In unloading this weapon, the magazine should first be withdrawn, and then the breechblock pulled to the rear sharply to eject the cartridge from the chamber.

Note that this arm differs from the standard Military Model pistol in having a magazine release stud provided on the right side of the

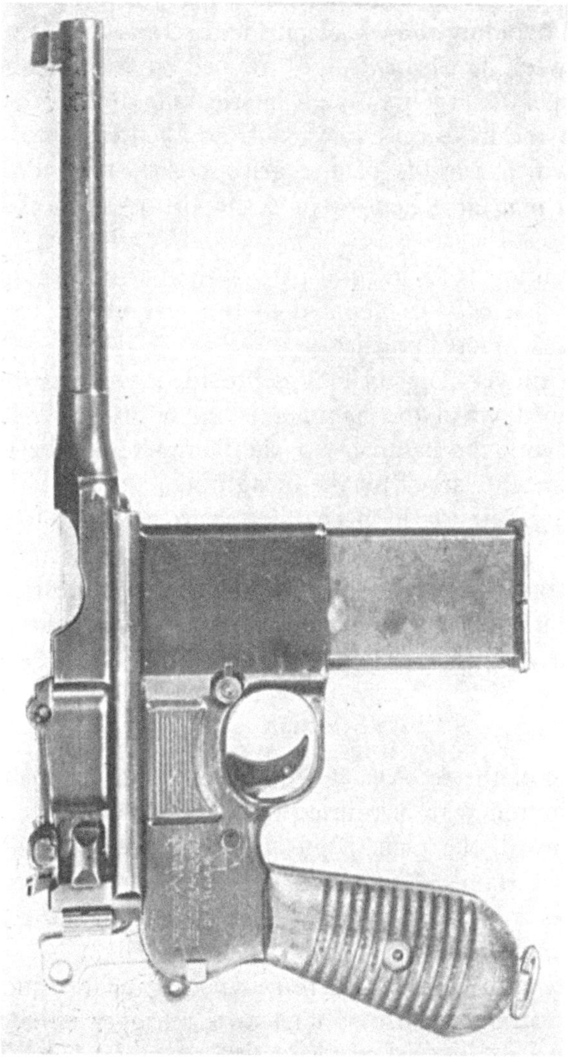

Mauser Model 712 semi- and full-automatic pistol with 20-shot detachable magazine inserted. Fire control switch is on left side of receiver. This arm is officially classed in the submachine gun category.

receiver. (The true pistol has a *fixed* instead of a *detachable* box magazine.)

Construction

This model follows all the general characteristics of the original military pistol in avoiding the use of pins and screws to retain working parts. The working elements are grouped in self-contained assemblies which operate largely through interlocking and through cam surfaces. All the mechanism is inserted from the rear into the receiver forging which consists of the grip section, the side plates, triggerguard and magazine housing, and supporting surfaces for the barrel extension.

The barrel extension is an integral part of the barrel forging itself. The breechblock assembly is mounted in the rear of the barrel extension in standard Mauser practice.

A new type of universal safety lock is provided which permits the safety to be applied when the hammer is up or down. When the safety is applied with the hammer up, the hammer may be lowered without danger since it cannot hit the firing pin.

Except for such differences in mechanism as are necessary to provide an automatic sear to enable full automatic fire, and the arrangement for detachable magazine insertion, the construction and design, while differing in weights of parts from the earlier semi-automatic military pistol, is identical with that earlier form.

Fire Selection

On the left side of the receiver to the rear of the trigger is an oval plate. A spring button with a knurled head forms the center of this plate. When forward, the plate points to the letter "N" indicating "Normal"—that is, standard semi-automatic fire in which one pull of the trigger fires a cartridge, ejects the empty and reloads ready for the next trigger pull.

When the plate is pushed to the rear it points to "R", indicating "Rapid"—that is, automatic fire as long as the trigger is held back.

This plate bears an integral pin passing across the inside of the receiver below the lockwork which has a bearing in the receiver.

Two cams are milled into this transverse plate pin, a large one near the center which engages with a spring-loaded lever which is pivoted to the trigger; and a smaller one on the right which is rounded to fit into a fork in an automatic sear disconnecting unit on the bottom of the receiver near the right hand sidewall.

Unlike the trigger in the standard type, this full automatic model trigger does *not* have a nose. A short spring lever bearing against the flat center cam on the transverse pin, is pinned directly to the trigger and is held against the cam by its spring. Since this lever is pivoted to the trigger below the trigger pivot line in the receiver, it provides long leverage.

While the loading, firing, locking and unlocking, extraction and ejection systems are about the same in both weapons, the action of the semi- and full automatic control requires attention.

Semi-Automatic Action

When the sliding plate on the left side of the receiver is pushed forward to "N", its pin inside the receiver brings the flat surface of its center cam to the trigger lever. Thus as the trigger is pulled it raises the sear lever and then slips forward out of it. This permits the sear to reengage in its bent in the tumbler as the hammer is driven back by the recoiling action after falling and firing the cartridge in the chamber. The action is substantially that of the standard Mauser Military Model pistol at this time.

Full Automatic Action

In full automatic action, the side plate is pushed back to point "R". In this position the pin inside is turned so that its cam faces the trigger lever, thereby advancing the trigger lever about .25 inch to the front. This changes the angle of trigger nose to sear lever; hence when the trigger is pulled, while the sear lever is *elevated* somewhat it still cannot slip off the nose of the trigger lever. Thus as long as finger pressure is maintained, the sear cannot engage with the hammer bent to hold the hammer back, and the arm will continue to fire as long as there are cartridges in the magazine.

While in some Spanish imitations of his pistol, the hammer follows the breechblock down to fire, this is not a good sear design. In the genuine Mauser *a secondary sear* is provided to make sure that the arm is completely locked before the firing of each round.

This secondary sear automatically holds the hammer back until the breechblock is fully locked, at which time it automatically releases.

It is positioned below and to the rear of the hammer pivot nearly opposite the primary sear. Like the primary sear, it has a broad bearing surface, reaching the entire width of the tumbler.

The secondary sear is a single unit having a wide bearing in each wall of the receiver. Its right side is shaped as a flat lever having unequal arms which position in recesses in the receiver, and it is flush with the primary sear spring and its lever. The lower arm is the shorter one. It engages with the disconnector bar. The long upper arm curves to the front where it ends in a rounded open hook projecting above the receiver into a slot in the underside of the barrel extension. This arm can move only within the limit provided by its recess. It is operated by a small coil spring and plunger mounted in a tunnel in the side of the receiver below the primary sear spring.

When the sear is in its notch, this upper arm projects above the level of the lock frame or receiver, and a forward movement will bring it below this lever. Thus as the breechblock travels to the rear and rolls over the hammer, the main bent is first caught by the primary sear and then by the secondary sear in its bent. As the secondary bent is cut further in the tumbler than the primary one, when the arm is firing full automatic, the hammer fall is somewhat longer as it is released by automatic or secondary sear action.

When the plate is set for full automatic fire, pressure on the trigger results in the primary sear being kept out of its bent as the hammer falls. As the breechblock during rearward movement unlocks from the receiver when the barrel is halted, the hammer tumbler after passing over the primary sear engages the secondary (automatic) sear momentarily.

Since the barrel extension is in full recoil at this point, the upper arm of the secondary sear thrust by its spring, rises up into its cut in the underside of the extension. The bent cannot engage until this point is reached.

The compressed recoil spring now reasserts itself and drives the breechblock forward to strip the top cartridge from the magazine and drive it toward the chamber. During the closing movement, the barrel extension begins this forward movement as the mainspring acts upon it through the rocker and the bolt lock. Thus the rear end of the slot in the underside of the barrel extension pushes the upper arm of the automatic or secondary sear, thereby releasing it from its bent, and permits the hammer to continue forward to hit the firing pin and fire the cartridge *when the breech is fully locked.*

The automatic disconnector during full automatic fire is unusual. There is a bar which can move in a groove in the bottom of the receiver. Its front connects with the transverse pin of the fire control lever, and its rear engages with the short bottom arm of the sear. When this bar is to the rear during normal fire, the secondary sear is held out of its bent and also out of its engagement with the barrel extension by this action.

The Safety

The safety pivot pierces the side of the receiver. A recess in the side of the hammer which receives this pivot to serve as a safety bar is cut out so that when the "safe" position is reached, the hammer can still be released by trigger pressure but will be halted in forward motion before it can hit the head of the firing pin.

On this design of universal safety, the sear is carried further to the rear on the left into a recess in the safety lever. The lever is cut to permit movement of the sear *only* at the fire and at the safe positions. It cannot function at any intermediate point. Thus when the hammer is up and the safety is applied, pulling the trigger cannot accidentally fire a cartridge, for as the hammer falls its fall is positively halted before hitting the firing pin.

Like all earlier Mauser military pistols, this arm is designed to be used with a stock shoulder holster. However it should be noted that the grip is larger and thicker than that of the standard Military Model pistol and the two stocks will *not* interchange.

While in theory this is an efficient police pistol, in actual practice the cartridges are discharged so rapidly and the muzzle elevates so quickly that it is not an efficient full automatic design. The full automatic feature is of value only in very unusual cases of emergency.

Characteristics

Following are the characteristics of this model 712:

The caliber is .30 (7.63mm Mauser). It uses the standard 7.63mm Mauser cartridge as already described.

The overall length of the pistol is 11.3 inches. The length of the *barrel complete with its chamber* is 5.2 inches. (It must be remembered in this design that the barrel *extension* which houses the breechblock is an integral part of the barrel forging, extending back from the chamber and being suitably cut away to permit recoil travel and also to house the breechblock assembly).

The weight of the arm is 2.79 pounds without the stock and with the 10-shot magazine inserted but empty. With the 20-shot magazine (empty) inserted it weighs 2.9 pounds.

The arm may be loaded from Mauser clips from above or by insertion of box magazine from below as already outlined.

While the rear sight is graduated to only 1000 meters, the cartridge in this pistol has an extreme range of about 2200 yards.

The muzzle velocity with standard barrel is about 1392 feet per second and the muzzle striking energy about 366 foot pounds.

At 50 meters the penetration is 8 to 9 inches in soft pine.

This weapon may be immediately identified by the Mauser trade mark on the left side of the receiver to the rear of the sliding fire control plate. It is manufactured of the finest materials to the finest standards of workmanship. It should *not* be confused with arms which look very much like it which were made in Spain; those are differently constructed mechanically and are neither as strong nor as reliable. The original Mauser Model 712 was first generally marketed in 1931.

32. MODEL 06-08, MAUSER AUTOMATIC PISTOL

At the time of its introduction this pistol was believed by Mauser and his associates to be a tremendous advance in the field of automatic weapons. The simple locking mechanism operating through check valves, permitted pistol manufacture at a much lower cost than had heretofore been possible in the Mauser factory. The original Mauser 96 pistol was an extremely expensive arm to machine.

However, the feature which most impressed Mauser's associates was, curiously enough, the elementary one of automatic loading by insertion of a loaded magazine when the arm was empty and in recoil position.

Under action of the magazine follower, the ejector was held back by a special catch, and when a loaded magazine was inserted, the catch was automatically disconnected. The breechblock was thereby automatically released and its spring drove it forward to strip the top cartridge from the magazine and to chamber it ready for firing. So important did they consider it, that the Mauser factory at the time of its announcement released the following statement in part: "This is therefore an innovation of utmost importance, completely fulfilling the basic requirement of all self-loading pieces—that after the firing of the last cartridge, the weapon must be ready for reloading, and immediately upon the insertion of a new, full magazine, it must at once be ready for continued firing. Domestic and foreign patent have been applied for to protect this method of the restoration of the readiness for firing in particular. In the meantime we must forego the reproduction and abstract at this place of these and other patents relating to this weapon for various reasons. However, the illustrations reproduced here should be sufficient to give a clear picture of the details of a design of this pistol too."

In actual practice, this strange weapon was one of the most clumsy and useless designs ever attempted. The arm is bulky, complicated to a high degree, and susceptible to many breakages and jamming causes not found in standard types of automatic pistols.

The one feature of interest in the arm is the valve or claw-type operation of locking and unlocking the breechblock, and even that feature is tremendously complicated in this particular arm.

This pistol was never manufactured in quantity. When it was introduced, the following statement was made about it by the factory:

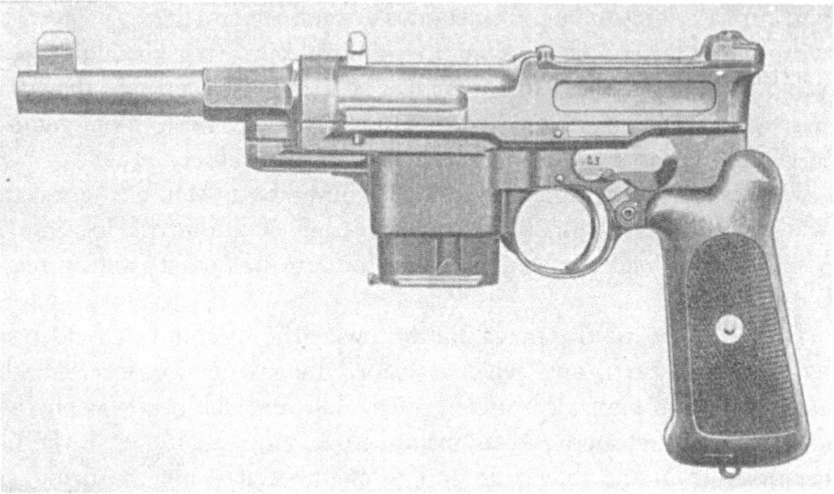

MODEL 06-08, MAUSER AUTOMATIC PISTOL, LEFT SIDE VIEW
This locked breech pistol utilizes the essential locking principle of the self-loading rifle 06-08. It is short recoil operated and locks through short turning flaps or valves.

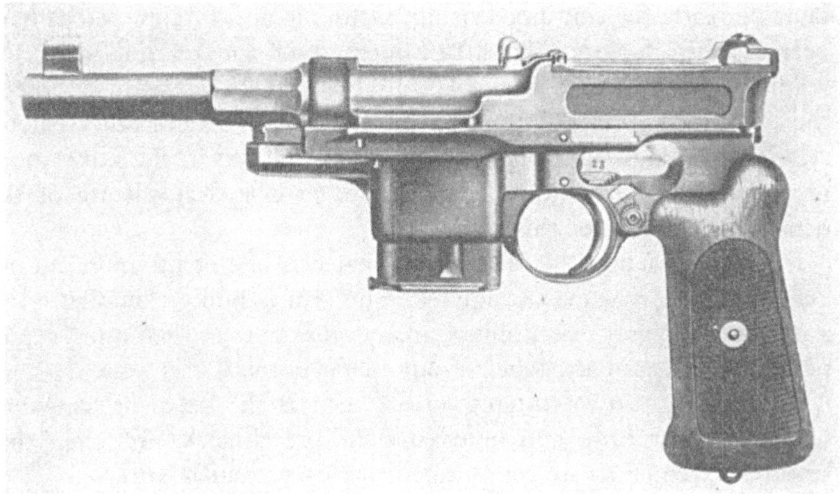

MODEL 06-08, MAUSER AUTOMATIC PISTOL. LEFT SIDE VIEW. IN FULL RECOIL POSITION WITH BREECH OPEN.

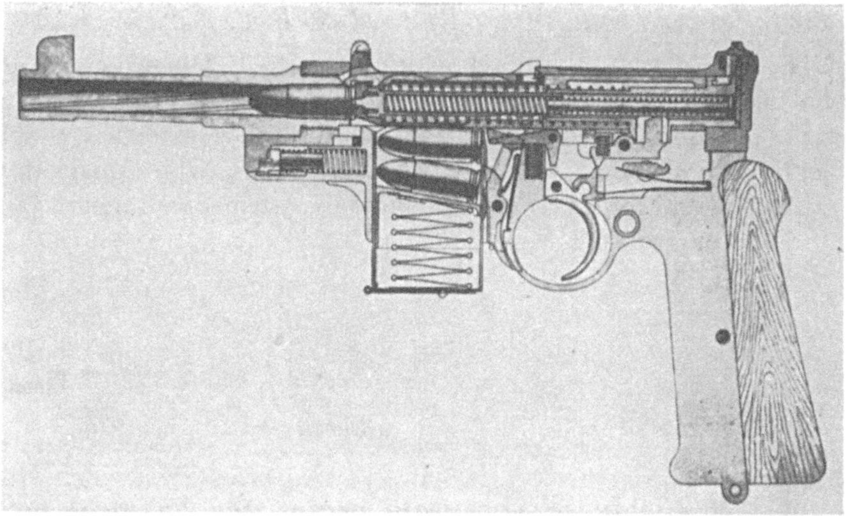

MODEL 06-08, MAUSER AUTOMATIC PISTOL, LEFT SIDE PHANTOM VIEW SHOWING STAGGERED MAGAZINE PARTLY LOADED AND A CARTRIDGE IN THE CHAMBER READY TO BE FIRED.

Note that in this design, a powerful separate *barrel return spring* is necessary and is mounted with its guide directly below the chamber in the forward part of the receiver ahead of the magazine.

This is a striker-fired weapon, the striker and its spring being mounted within the breechblock. The firing pin return spring is at the point of the pin itself. The recoil spring is also mounted in the breechblock. The trigger and sear are hooked up with the sear release ready for pressure on the trigger.

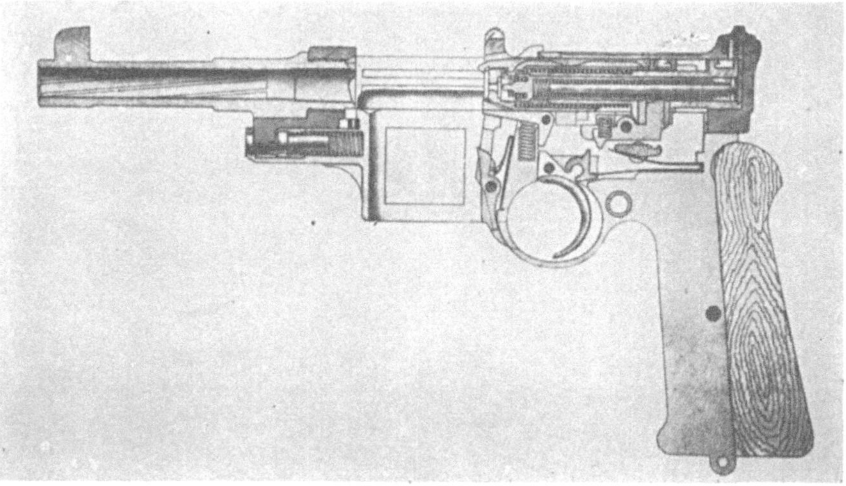

MODEL 06-08, MAUSER AUTOMATIC PISTOL, LEFT SIDE PHANTOM VIEW SHOWING PISTOL IN FULL RECOIL POSITION

When the last cartridge has been fired, the rear of the magazine platform rises to intercept the bolt or breechblock and hold it in rear position. Note that the sear cannot make engagement at this point.

"The breech lock in this novel self-loading pistol is designed essentially on the principle utilized in the self-loading rifle C. 06-08 which we have just discussed. Thus this pistol too, has that extremely simple locking mechanism with the two check-valves, which make this weapon definitely superior to all previous systems, for military purposes in particular."

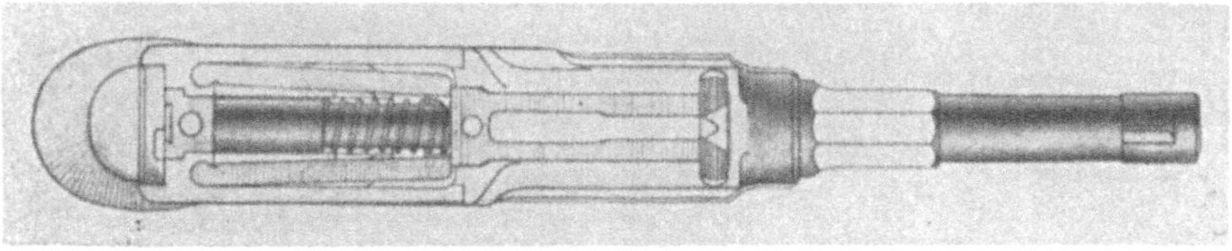

MODEL 06-08, MAUSER AUTOMATIC PISTOL, TOP PHANTOM VIEW SHOWING POSITION OF LOCKING LUGS AND METHOD OF CAMMING THEM OUT AND IN TO UNLOCK AND LOCK

This system follows very closely that of the Model 06-08 automatic rifle developed by Mauser.

While this pistol was not successful commercially, it was actually manufactured in testing quantity.

It is included here because of that fact.

It is not practical even to list the hundreds of hand built Mauser experimental arms; but every production model is included in this book.

33. MAUSER VEST POCKET PISTOL, CALIBER 6.35MM BROWNING (.25 AUTOMATIC COLT PISTOL) (W.T.P. I)

In the course of its manufacturing history, Mauser produced two types of vest pocket pistols.

The first of these arms, known as the W.T.P. 1 (Westentaschen Model 1), was introduced to meet the demand in Germany for a native vest pocket pistol to fire the 6.35mm (.25 Automatic Colt Pistol) cartridge. It was designed to compare in size with the Colt Vest Pocket Automatic introduced in 1908 and its Belgian counterpart, the Baby Browning.

The Mauser W.T.P. 1

This W.T. 1 Model was continued in manufacture until 1939.

It is a standard blowback type of weapon, the low powered cartridge employed not requiring the use of a locking system other than that provided by the weight of the moving parts and the spring to be compressed. (In Germany this type is called "Federverschluss"—literally "spring locked").

This pistol will use the standard United States caliber .25 Automatic Colt Pistol cartridge. It weighs only 11-ounces, is 4.5 inches overall in length, has a barrel 2.4 inches long, a maximum height of 3 inches, and is .75 inch thick at its widest point. The magazine holds 6 cartridges. By loading another cartridge in the firing chamber directly, the capacity is raised to 7 cartridges.

In general exterior appearance, it somewhat resembles the Colt design. However, it is *not* equipped with a grip safety as is the Colt.

Loading

The arm is loaded by pushing back the magazine catch in the bottom of the butt at the rear, which releases the standard steel box magazine to be withdrawn through the bottom of the grip section of the receiver. The magazine is loaded by pressing the first cartridge down on the front end of the magazine platform (or follower) and pushing it back under the overhanging lips of the magazine. Succeeding cartridges are started by forcing each one down on top of the one below it until the cartridge column is depressed enough that the cartridge may be slid back under the lips which overhang part of the top of the magazine to retain it. The loaded magazine is then inserted in the butt and pushed until it locks.

The serrated sections at the rear of the slide are then gripped by the thumb and fingers of the left hand while the pistol is held in the right hand, care being taken to see that the firing finger is *not* inside the triggerguard. Drawing the slide back to its fullest extent compresses the recoil spring mounted about its guide in the receiver channel below the barrel, to provide energy for the return movement of the slide.

As the slide is drawn back, it pulls with it the striker which is mounted in the rear breechblock-section of the slide in line with the chamber. The spiral spring around the striker is compressed as the striker is withdrawn. The sear catches in the striker bent and holds the cocked striker in rear position. The spring in the magazine thrusting up against the magazine follower pushes the cartridges up, and as the solid breechblock section of the slide passes over the top of the magazine, the top cartridge can rise far enough to be in line with the breechblock when that unit goes forward.

Releasing the grip on the slide permits the recoil spring to pull the slide forward, and the feeding face on the breechblock, striking the top of the cartridge case in line, thrusts it forward out of the retaining lips of the magazine, and into the firing chamber. The extractor in the breechblock snaps over the cartridge case and engages in its extracting groove. The pistol is now ready to fire. There is a thumbpiece on the side of the receiver on the left which may be pushed up to lock the action so that the trigger cannot release the striker. Unless this thumb safety is applied, the pistol is now dangerous. When the striker is cocked, its head protrudes through the rear of the slide. Thus, if a protrusion can be felt or seen, one is warned that the arm is cocked, though the chamber is *not necessarily loaded.*

With the safety off (in the "down" position) pressing the trigger will force back the trigger mechanism to free the striker pin and let its spring drive it forward to fire the cartridge in the chamber.

Operation

As the cartridge is fired, the bullet travels down the barrel and the side pressure within the cartridge case forces the elastic brass case tightly against the walls of the firing chamber to act as a gas seal, while the *rearward* thrust of the gas within the cartridge case is transmitted to the face of the breechblock. Since the bullet is very much lighter than the breech assembly, it is out of the barrel before the rearward action has opened the breech appreciably. As the breechblock is blown back, the pressure within the cartridge case drops rapidly, and the cartridge case contracts. Thus the extractor in the fact of the breechblock draws the empty cartridge case back with it until it is ejected through a port in the slide. As the slide goes back, it operates an interrupter mechanism which prevents the trigger from functioning until ready for the next shot. The breechblock during its rearward stroke, compresses the recoil spring below the barrel and also the spring around the striker. The sear catches the striker and holds it back as the recoil spring pulls the slide assembly forward to chamber the next cartridge rising from the magazine.

Before another shot can be fired, pressure on the trigger must be released to enable the trigger spring to push the trigger forward far enough to operate the interrupter so that contact can be made for the next pressure on the trigger.

A pocket pistol which would fire automatically as long as the trigger was held back would be an extremely dangerous instrument, as the pistol would rise rapidly after each successive recoil. This interrupter mechanism assures that only one shot will be fired with each pull of the trigger. It is possible for a pistol expert to fire about 6 shots in one second, however, with this arm. When the last shot has been fired, the magazine follower rises to interfere with the forward movement of the breechblock, thereby holding the breech open to inform the shooter that his pistol is empty. When the magazine is withdrawn, the slide runs forward. When a loaded magazine is inserted in the handle from below, the slide *must again be drawn back*, and released in order to chamber the first cartridge to start firing again.

Cartridge Specifications

The cartridge used in this weapon is the standard United States .25 Automatic Colt Pistol cartridge. This cartridge is known in Europe and South America as the 6.35mm Browning cartridge. It uses a powder charge of about 1.5 grains of smokeless and fires a 50-grain metal jacketed lead bullet. It develops a muzzle velocity of about 755 feet per second, has a striking energy at the muzzle of about 71 foot pounds and at 15 feet will penetrate 3 pine boards, ⅞-inch thick.

While this pistol is not considered an effective side arm in the United States, it has tremendous vogue throughout the rest of the world as a personal protection arm. It is the second most popular pistol caliber in the world, the first being the .32 Automatic Colt Pistol cartridge, known in Europe as the 7.65mm Browning cartridge.

34. MAUSER VEST POCKET PISTOL, CALIBER 6.35MM BROWNING (.25 AUTOMATIC COLT PISTOL) (W.T.P. II)

This weapon was introduced in 1939 by Mauser as an improved form of the earlier Model I. It can be distinguished at once from the earlier Model by the fact that its grip is much more sharply curved, extending back over the top of the hand, and by the position of its thumb safety. The thumb safety on this model operates below the left hand stock, the thumbpiece protruding directly to the rear of the trigger. When forced down it leaves the pistol ready to fire. The general mechanism is the same as the earlier W.T.P.I. The slide in this new model bears the stamp "W.T.P.—6.35—D.R.G." (The earlier Model I bears on the slide "T.—6.35").

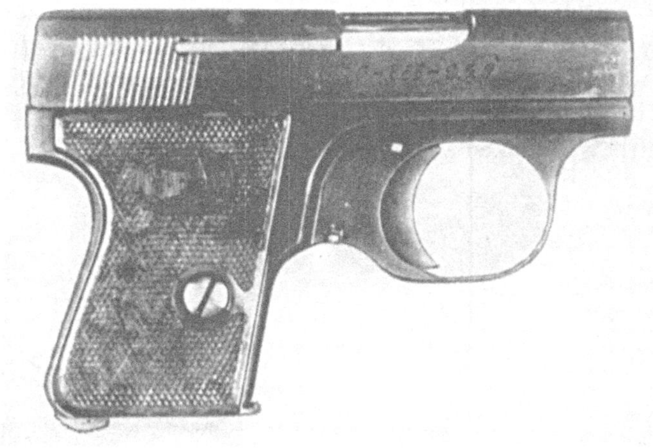

The Mauser W.T.P. 2

Because of the shift in the safety position, the new model has higher stocks which reach to just below the slide and are fastened with screws.

The dimensions of the Model II are as follows: Overall length 4.06 inches, barrel length 2.02 inches, maximum height 2.75 inches, weight 9.5 ounces. The magazine capacity is 6 cartridges, the same as in the earlier model.

This arm was designed specifically to fit better in the shooting hand than did the earlier type and to weigh about 2-ounces less. In workmanship, material and finish this is an excellent weapon.

though it does not have the complete safety factors embodied in the Colt Automatic Pistol, which is provided with a grip safety which functions automatically.

Both models of this Mauser W.T.P. are equipped with magazine removal safety. When the magazine is withdrawn from the grip it is impossible to pull the trigger. This feature was introduced to prevent accidents, which sometimes happen with pistols not so equipped, when shooters unfamiliar with semi-automatic weapons remove the magazine without checking the firing chamber. In an automatic pistol, not only must the magazine be removed to be emptied, but the slide must be pulled back to eject the cartridge (if any) in the firing chamber before the pistol is truly safe *unless* it is equipped with a magazine safety.

So long as there are any cartridges in the magazine, every time an automatic pistol is fired the arm is re-cocked and the chamber reloaded by the recoil. Hence, if the magazine is withdrawn and the pistol is not equipped with the special magazine safety, a pull on the trigger will fire the cartridge in the chamber. This chamber cartridge is frequently overlooked even by experts.

35. MAUSER AUTOMATIC POCKET PISTOL, CALIBERS 6.35MM AND 7.65MM BROWNING (.25 AND .32 AUTOMATIC COLT PISTOL)

(Taschen-Pistole mit Federverschluss)

Mauser pocket pistols were manufactured in only two calibers for general production. The first, introduced in 1910, was for the 6.35mm Browning cartridge which we know as the .25 Automatic Colt Pistol cartridge. It was followed by an arm of 7.65mm caliber of the same design, differing only in dimensions as required by the increased caliber and by the finger serrations on the slide. In 1934, a new model (Neues Modell) was introduced which varied mechanically very

The Mauser 6.35 mm (.25 ACP) Pocket Model

little from the original type. In this new model, however, the grip was narrower and instead of stocks screwed to each side of the grip frame, a single shaped grip of wood which covered both sides and the back strap of the frame was used. This grip was worked over the back of the steel grip section of the receiver, thrust into place, and then fastened with a stock screw.

Because of variable types of grips used, this pistol as encountered may seem to differ from the one pictured here. This illusion is produced by the fact that differently shaped grips of wood could be obtained to provide a different pitch to suit the hand of an individual shooter.

The dimensions and characteristics of the .25 caliber are as follows: Overall length, 5.25 inches, barrel length 2.96 inches, height 4 inches, thickness 15-16th inch, weight, 15 ounces, magazine capacity 9 cartridges.

The dimensions and characteristics of the .32 caliber are as follows: Overall length, 6 inches, barrel length 3.39 inches, height 4.5 inches, thickness 1.12 inches, weight, 21 ounces, magazine capacity 8 cartridges.

Characteristics of the .32 ACP Cartridge

The characteristics of the .25 Automatic Pistol cartridge have been given in Chapter 30. Following are the characteristics for the .32 Automatic Colt Pistol cartridge, known in Europe as the 7.65mm Browning: This cartridge shoots a lead bullet with a full metal jacket weighing about 74 grains at a muzzle velocity of approximately 965

The Mauser 7.65 mm (.32 ACP) Pocket Model

feet per second. The striking energy at the muzzle is about 152 foot pounds. The bullet has a penetration in ⅞ inch board of soft pine of 3 boards at a distance of 15 feet from the muzzle.

The Pistols

These pistols are customarily of fine material, workmanship and construction. They are equipped with a thumb operated safety but do *not* have the automatic grip safety found in the Colt type. These are striker operated weapons. A spring housed within the striker pin is compressed when ready to fire. Upon release by the sear the spring drives the striker pin forward to fire the cartridge. When the striker is cocked, the head protrudes through a slot or hole in the rear of the slide. Thus, when this pin can be felt or seen, the striker is cocked.

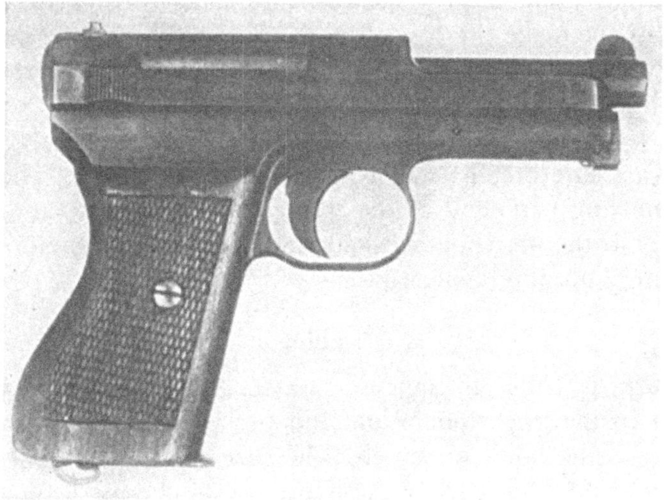

The New Model (Neues Modell) 7.65 mm (.32 ACP) Pocket Pistol

Note, however, that this does not necessarily mean that there is a cartridge in the chamber, as this is a *cocking indicator* and *not a cartridge indicator*. This good feature also has certain drawbacks. If the pistol is dropped and hits on the striker-projection, it is possible at times for the striker to be jarred off the sear to fire the chamber-cartridge.

A magazine safety is also an integral feature of these arms. When the magazine is withdrawn, the weapon cannot be fired. This prevents danger from a loaded cartridge remaining in the chamber when the magazine has been withdrawn. An automatic interceptor functions when the slide recoils to break the connection between the trigger and the sear, thereby preventing the firing of more than one shot for each pull of the trigger. The thumb safety is of unusual design, and consists of two parts. A thumb lever on the left side of the grip behind the trigger is pushed down to positively prevent firing. This also locks the slide so that the breech cannot be opened. Pressing the spring-controlled button directly below the thumb lever releases the lever and sets the pistol ready for firing.

When the last cartridge has been fired, the slide is held open. Inserting a loaded magazine, then drawing back slightly on the slide will release it, and allow it to go forward under the pull of the recoil spring to chamber a cartridge and leave the weapon ready to resume firing. The substantial grip, good balance, absence of projecting parts and flat shape make this a good pocket pistol. This model belongs tc the blowback category listed by the Germans as "Mit Federverschluss" or "with spring lock."

Because of the comparatively low power of the cartridges employed, a secure breech locking system is not necessary. The weight of the moving parts and of the springs resisting the opening of the breech hold the breech closed long enough to permit the pressure to drop before opening appreciably.

Loading

The arm is loaded in standard fashion. A loaded box magazine is inserted in the grip from below and pushed in until the magazine catch locks and holds it securely. The slide in this arm is of unusual design and does not enclose the entire top of the barrel in common pocket automatic pistol fashion. However, it is machined at the rear to provide the customary gripping surfaces, and when this breech-block-slide is drawn back it cocks the striker and permits a cartridge to rise in the magazine. When the grip on the slide is released, the recoil spring housed below the barrel acts to draw the slide forward to chamber a cartridge and leave the weapon ready for firing.

Barrel and Trigger Design

The barrel design is unusual in that it carries the front sight at its forward end, while its underside is provided with projections to permit locking it firmly to the receiver as the rear projection is worked down into a slot in the receiver, and the recoil guide spring rod inserted from below the muzzle passes through a ring under the barrel in the front projection.

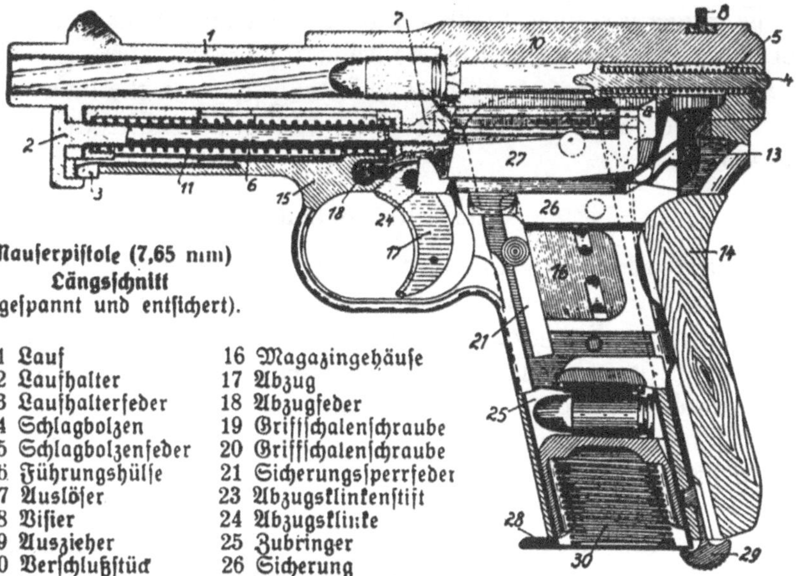

Mauserpistole (7,65 mm)
Längsschnitt
(gespannt und entsichert).

1 Lauf	16 Magazingehäuse
2 Laufhalter	17 Abzug
3 Laufhalterfeder	18 Abzugfeder
4 Schlagbolzen	19 Griffschalenschraube
5 Schlagbolzenfeder	20 Griffschalenschraube
6 Führungshülse	21 Sicherungssperrfeder
7 Auslöser	23 Abzugsklinkenstift
8 Visier	24 Abzugsklinke
9 Auszieher	25 Zubringer
10 Verschlußstück	26 Sicherung
11 Vorholfeder	27 Abzugsstollenhebel
12 Auswerfer	28 Magazinboden
13 Doppelfeder	29 Magazinhalter
14 Griffschale	30 Zubringerfeder
15 Griffkasten	31 Deckplatte

a) Abgesetzte Kante des Abzugsstollenhebels (27).
b) Nase des Schlagbolzens.
c) Stoßboden.
d) Teil des Auslösers (7), der in die Laufbahn (Kammerbahn) tritt.
e) Sicherungshebel.
f) Drehpunkt des Sicherungshebels (-stücks).
h) Knopf der Sicherungssperrfeder (21).
i) Fenster zum Auswerfen der Hülse.

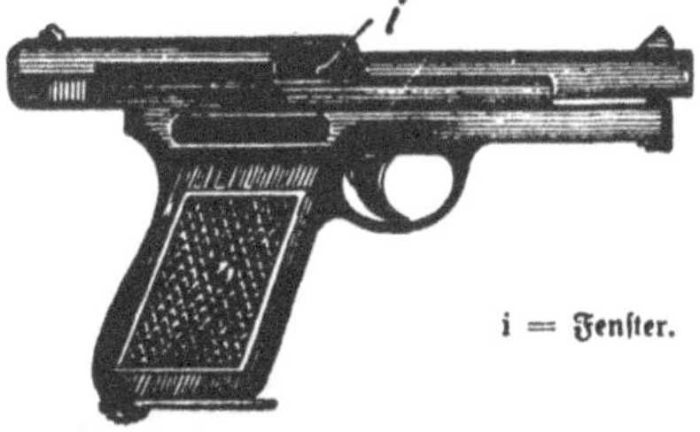

i = Fenster.

Verschlußstück nach dem Abfeuern in hinterster Stellung.

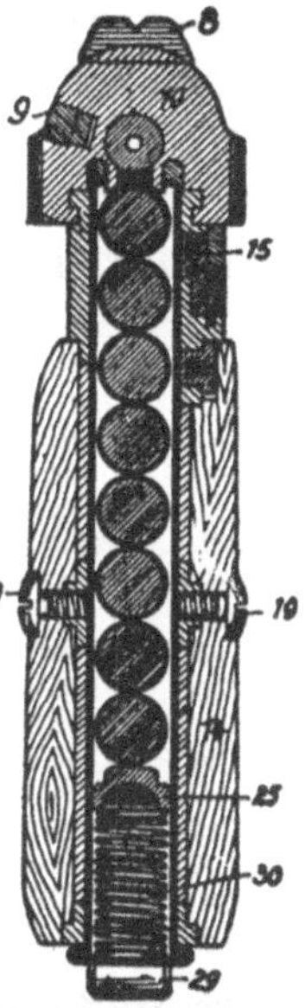

Querschnitt durch Verschluß und Griffstück
(von vorn gesehen).

MAUSER POCKET AUTOMATIC PISTOL CAL. 32 (ALSO MADE IN CAL. 25 AUTO) TOP LEFT SIDE VIEW WITH MAGAZINE AND CHAMBER LOADED. THE PISTOL IS READY TO FIRE.

Right. Rear view showing arrangement of cartridges in straightline magazine. Left. Right side view at end of recoil stroke showing ejection port in slide.

The trigger is equipped with a heavy boss having a thick pin milled out of the receiver on the left side below the breech. There is a nose above and to the rear of this boss which engages with the interceptor block placed above it. This interceptor block is pivoted to the tail of the sear.

Interceptor Action

A small coil spring below the interceptor acts to thrust the upper face of the interceptor through a groove in the receiver extending up into a corresponding groove in the underside of the slide. When the slide recoils, its solid surface hits this interceptor face and drives it down inside the receiver. In this position, the interceptor is drawn out of contact with the trigger nose. Hence, while the sear can catch and hold the striker in rearward position, the trigger cannot come into contact with the sear lever to release the firing mechanism until the slide is fully forward and closed and the trigger is released to permit its spring to push it forward so that the interceptor (or interruptor) block can rise under influence of its spring into the cut on the underside of the slide. The sear is a long lever operating on a pivot. Its rear upper end is designed to catch in a slot in the underside of the striker as the striker is forced back during rearward motion of the breechblock-slide.

Operation

When the trigger is pressed, it rotates on its pin and a small catch at its upper rear is elevated to hook under the edge of the sear lever. This depresses the rear of the pivoted lever, meanwhile compressing the sear spring below, and draws the sear out of contact with the striker.

The compressed striker spring within the striker tube thrusts the striker forward to hit the cartridge in the chamber and discharge it. As the bullet travels down the barrel, the slide starts to move much more slowly to the rear. The extractor in the face of the breechblock end of the slide carries the empty cartridge case back with it. The forward arms of the slide below the barrel carry the recoil spring back and compress it around its guide below the barrel. As the slide opens, exposing the breech opening for ejection, a solid surface on the slide on its under travel side on the left runs over the head of the interceptor (or interruptor) block and pushes it down. This forces the trigger catch below it down to a point where it cannot engage with the sear lever. The sear spring thrusting the rear end of the sear lever up, enables it to engage in the underside of the striker cut as the striker reaches full compressed position.

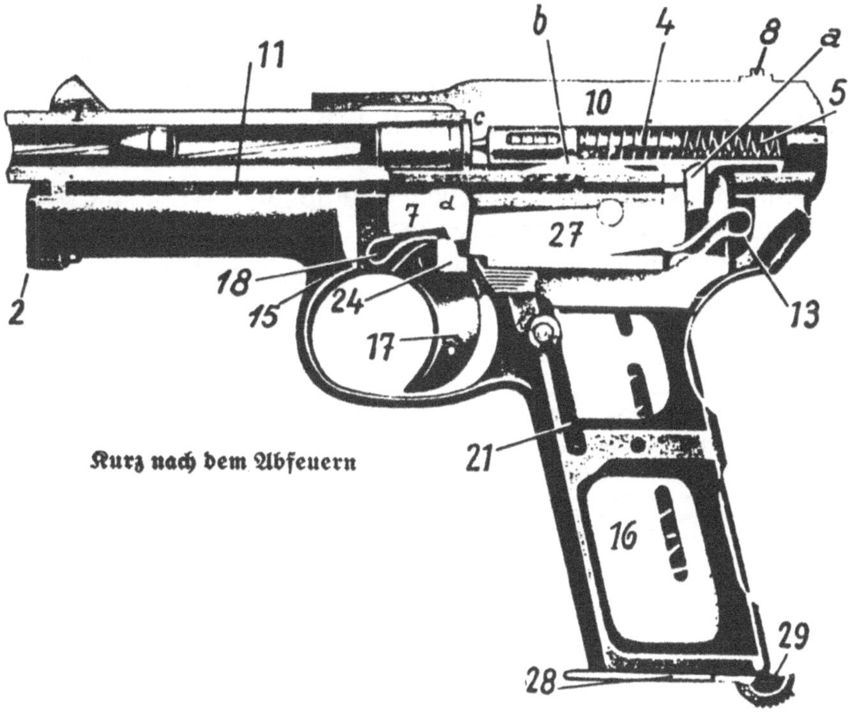

MAUSER POCKET AUTOMATIC PISTOL CAL. 32 (ALSO MADE IN CAL. 25 AUTO) LEFT SIDE VIEW AT INSTANT OF FIRING. BULLET IS STILL IN BARREL

At the end of the recoil stroke, after the empty cartridge case has hit the ejector and been hurled out of the pistol, the magazine spring forces a cartridge up into line. The recoil spring, now fully compressed, reasserts itself and drives the slide forward to chamber the top cartridge from the magazine. When the trigger is released momentarily, its spring moves it forward and also thrusts the interceptor block up into its cut in the underside of the slide. At this point, the trigger catch is again close to the sear lever and another pull on the trigger will fire the next shot.

Safety Design

The *safety* is of unusual design. It has a spring operated arm pivoted to the receiver. When the thumbpiece at the forward end of this arm is pulled down, the rear end of the safety bar rises into contact with the rear end of the sear, making it impossible for the sear to release the striker under any condition. Furthermore, a tip on the end of the safety is thrust up to double lock the striker. Both the sear and the striker are thus solidly locked. When the spring operated button below the thumb lever is pressed, the double action sear spring and the flat safety spring function to thrust the rear of the safety lever or bar down out of engagement and thereby pivot the front thumbpiece up ready to be applied again when necessary.

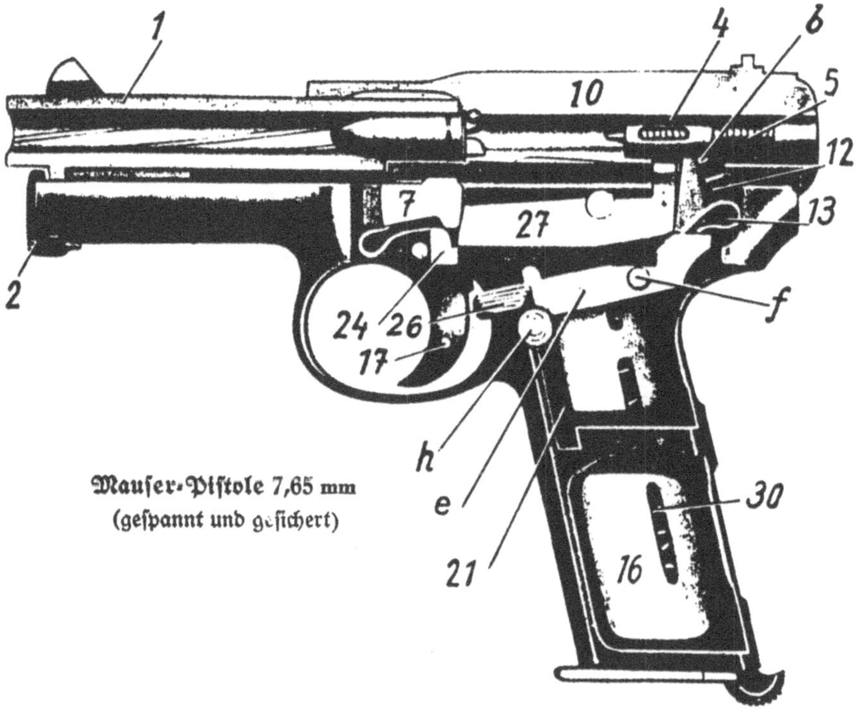

MAUSER POCKET AUTOMATIC PISTOL CAL. 32 (ALSO MADE IN CAL. 25 AUTO) LEFT SIDE VIEW WITH CHAMBER LOADED BUT SAFETY APPLIED.

This pistol is an essentially simple design. Like the Mauser Military Pistol, it has a maximum of built-in contacts and bearing surfaces and does not depend upon pins or screws for its assembly.

Stripping

Dismounting this arm is quite simple: The recoil spring guide protruding with its locking edges from the receiver below the muzzle, is pressed and turned to free it. It is then pulled forward for removal. The slide is then drawn back over an empty magazine. The barrel may then be lifted straight up out of the receiver. Inserting a magazine and then withdrawing it will release the slide and permit the recoil spring to pull it forward. The slide and recoil spring assembly will come off the receiver runners to the front. The striker and extractor may be removed from the slide. Removing the stock screws and working the stock off the back of the receiver grip, then working off the receiver lock plate on the left side will expose all the action with its interlocking parts for removal if necessary.

PARTS KEY TO DRAWINGS ON 207-208

1. Barrel
2. Barrel Retainer
4. Striker
5. Striker Spring
7. Interceptor
10. Slide
11. Recoil Spring
12. Ejector
13. Double Spring
16. Magazine
17. Trigger
21. Safety Lock Spring
24. Trigger Catch
27. Sear
e. Thumb Safety Bar
f. Safety Pivot Pin
h. Safety Release Button

36. MODEL H Sc MAUSER CALIBER .32 AUTOMATIC PISTOL

This is the last production automatic pistol designed by Mauser engineers before World War II halted their activities.

This arm uses the standard .32 Automatic Colt Pistol cartridge. The magazine holds 8 cartridges and another cartridge may be inserted in the chamber giving a capacity of 9.

The barrel is 3.37 inches long, the overall length is 6.5 inches, and the pistol weighs 20.6 ounces. The action is straight blowback, no lock being necessary in a weapon of this caliber and design.

This pistol incorporates a double action feature of general revolver type. Under this system, it is possible to carry the hammer down on a loaded chamber and to bring the pistol into rapid action by merely pulling straight back on the trigger. Through lever and spring arrangement, the hammer will be rotated to the rear and then slipped to hit the firing pin and discharge the cartridge in the chamber. The blowback or recoil will then force the slide back and cock the hammer for the succeeding shots in standard automatic pistol fashion. As in all automatic pistols, however, the firing chamber must be manually loaded for the first shot.

In all earlier forms of Mauser pocket pistols, whenever the chamber is loaded, the striker mechanism *is at full cock*. If an arm is carried this way, it is always potentially dangerous unless the thumb safety is kept applied. In periods of emergency, time may be lost in remembering to push the safety off. Another bad factor of this type of design is that if the weapon is carried ready for action, the striker spring is at full compression and may be weakened over a period of time.

With the double action feature, it is possible to carry the chamber loaded in complete safety with the assurance that the mainspring, not being under compression, will not be weakened. When it is desired to fire the pistol speedily, there is no necessity to think of safeties (though a thumb safety is provided to give double assurance to those desiring it). A pull on the trigger, as already stated, will fire the weapon without attention to any levers or buttons.

While the essential feature of this new Mauser pistol is the double action cocking system, it also represents a forward step in pistol development in the streamlining of design, in the shape and pitch

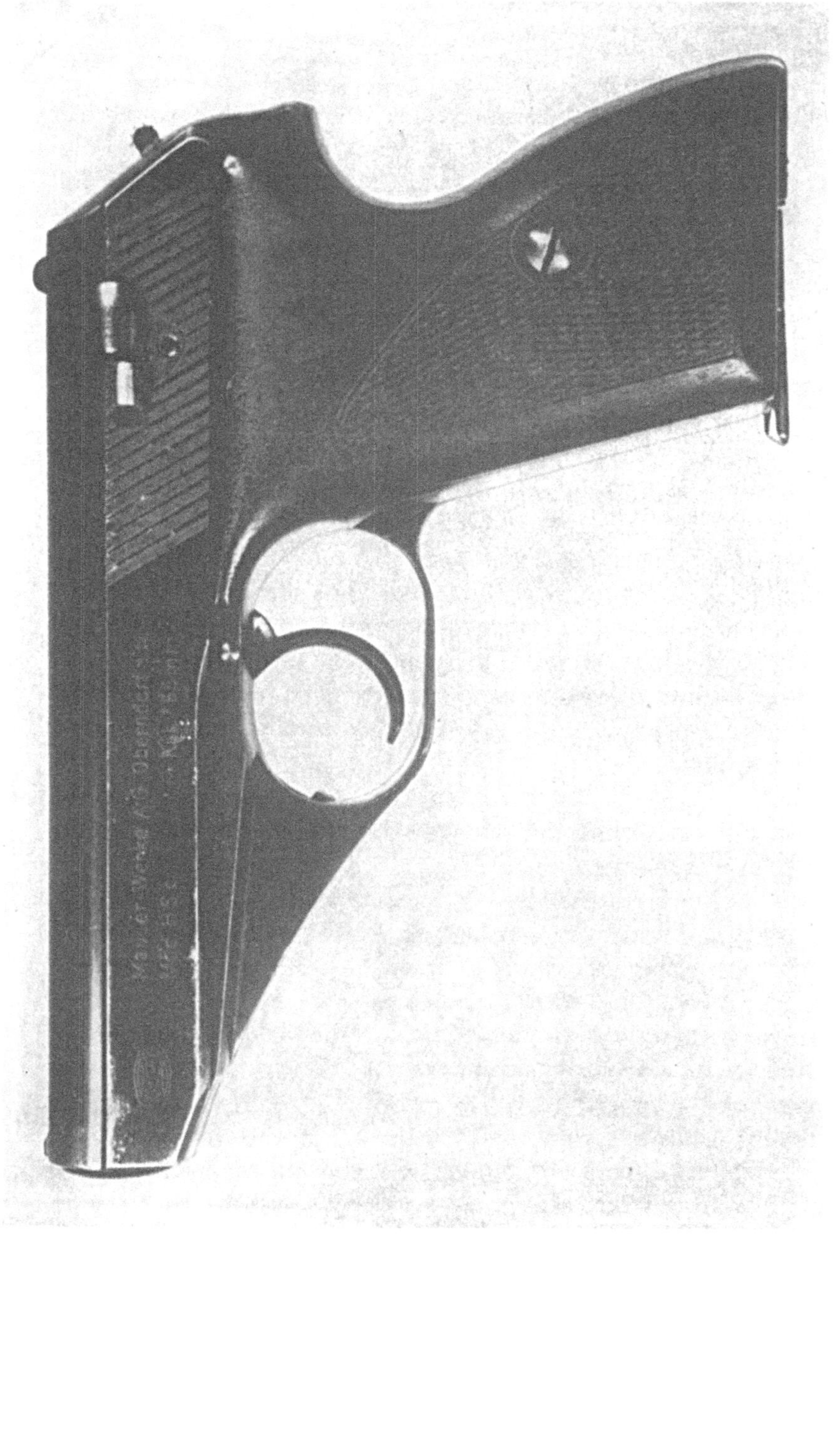

Mauser 7.65 mm (Cal. 32 ACP) H Sc.
Small projection to rear of slide is exposed tip of hammer, which is shown cocked.
Hammer may be safely lowered on loaded chamber, and first shot fired by a double-action pull as in a revolver.

of the grip which permits more instinctive pointing, and in the simplified takedown system.

Parts of the Weapon

The principal parts of the arm are: The barrel, the slide whose rear section is the breechblock, the receiver which forms the grip and also houses the firing mechanism, the sights, the trigger mechanism and the safety. With the sole exception of the grips all parts are made of steel burnished for protection against rust. The grips are usually of plastic but may be of wood. (Note: Pistols made during the War were often poorly finished without burnishing).

The Barrel. The barrel forging is unusual. It is rifled with 6-grooves to the right. The rear chamber section is heavily reinforced and provided with an abutment against which the rear of the recoil spring, which is mounted concentrically around the barrel, is compressed. Two slanting teeth are provided on the underside of the chamber section of the barrel to lock securely into the receiver cuts prepared for them and provide a secure and positive mounting for the barrel.

The Slide and Breech Lock Mechanism. The slide incloses the barrel for its entire length. Its rear section serves as the breechblock, and it is suitably machined to travel back and forth in the receiver guides in a straight line. At its front end where it surrounds the barrel muzzle, it is machined to provide a seating for the forward end of the recoil spring.

In the breech section are housed the firing pin and its spring, the extractor and its spring, the safety, and a buffer pin each for the extractor and the safety.

When the pistol is assembled, a locking catch, spring controlled, which may be released by pressure on its projection in the forward end of the triggerguard, serves to keep the teeth on the underside of the barrel firmly locked to the receiver.

The firing pin and its concentrically mounted spring are of standard design; the spring serving to pull the firing pin back into the breechblock immediately after the firing pin (driven forward by the hammer) hits the primer of the cartridge in the chamber to discharge it. Thus as the action opens, the firing pin cannot be injured by the rearward pressure of the cartridge case being thrust back against the face of the breechblock, a thrust which is quite forceful in a blowback pistol.

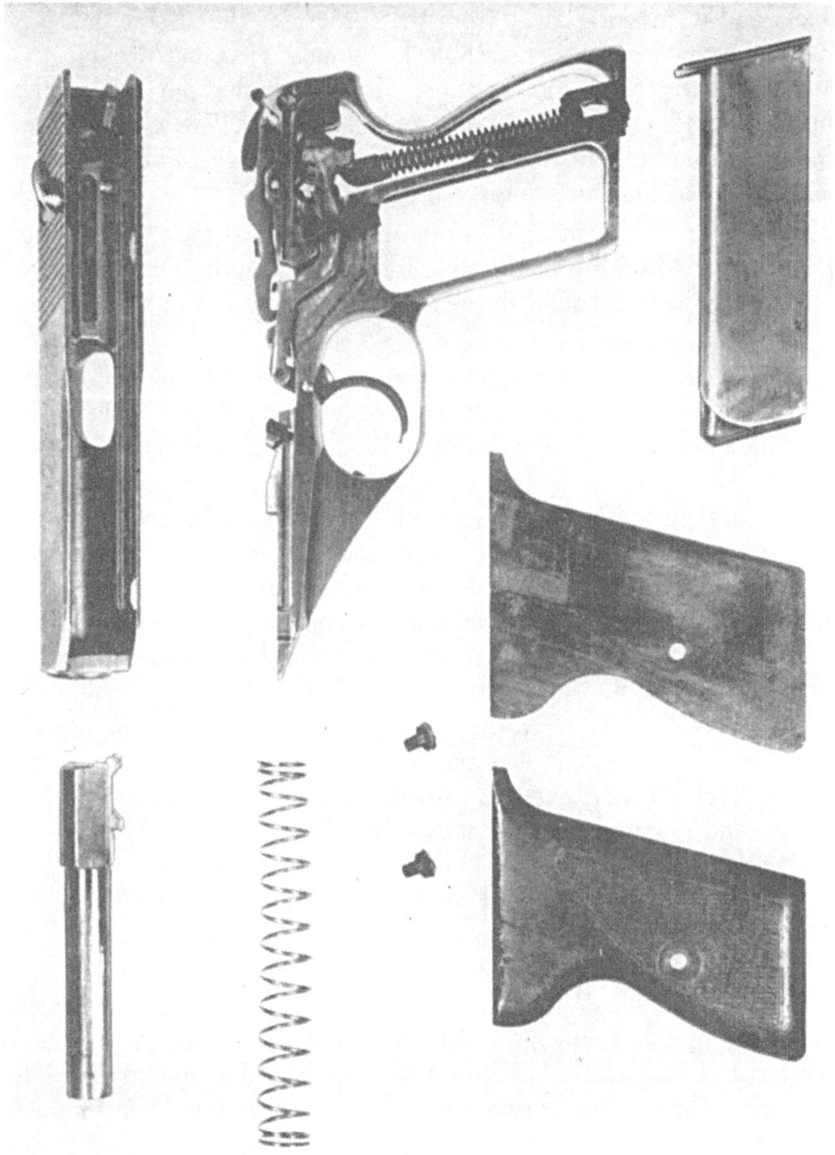

Mauser 7.65 mm H Sc.
Top: Barrel and slide. Middle: Recoil spring; receiver with firing mechanism. Bottom: Stocks and magazine.

Buffer pins inserted into the extractor spring activate the extractor under pressure of its spring both when snapping into the extracting groove in the cartridge case on forward movement of the breechblock, and in pivoting away from that groove as the ejector on rearward movement of the breechblock compels ejection through the port in the right side of the slide. The rear buffer pin retains the thumb safety in the slide in its applied position. These assemblies should not normally be dismounted. They are intended to be removed *only for repair* and by a competent gunsmith.

The Sights. The front sight in this arm is a part of the slide forging. A rib is matted on top of the slide for its full length. The rear sight is dovetailed into a cut in the rear section of the top of the breechblock slide.

The Receiver. The receiver forms the handle of the arm, guides the travel of the breechblock back and forth during recoil and houses the following parts: The rotating hammer, with its strut and mainspring assemblies and with pin to the sear lever; the trigger with its spring and pin; the trigger sear, the interruptor, the trigger catch with its spring and pin, the angle piece, sear lever, mainspring, ejector with spring, barrel holder with spring, grip plates and screws, magazine complete, and magazine catch and pin.

Safety. The safety in this weapon is a thumb lever mounted on the left hand side of the slide at the rear with a shaft passing inside the slide. On rotating the lever, the firing pin is pulled within the breechblock and its rear is elevated and locked so that the falling hammer cannot reach its head to fire a cartridge accidentally; and so that any straightline pin movement is impossible.

The Magazine Assembly. This consists of the standard steel box type magazine with sides cut away enough to permit counting the cartridges it contains. The magazine box has top lips which are folded over to permit insertion of cartridges from the forward end when they are thrust down on the follower and then slid in underneath the lips, and which are cut away at the rear enough to permit the head of the breechblock to hit the top cartridge and strip it into the firing chamber on forward movement of the slide. The follower or platform on which the cartridges rest, the magazine spring below the follower, and the bottom plate which serves as the lower compression point for the spring are the other components.

The magazine catch is a thumbpiece in the rear of the lower part of the grip which is pushed back to release the magazine for withdrawal from below. This catch operates under pressure from the mainspring which lies above it in the rear of the grip section of the receiver, and which thus provides two spring functions.

The Coordinated Action of the Parts

When the safety is off, and a loaded magazine is inserted in the grip, pulling back on the slide will cause the forward section of the slide around the muzzle to thrust back the recoil spring around the

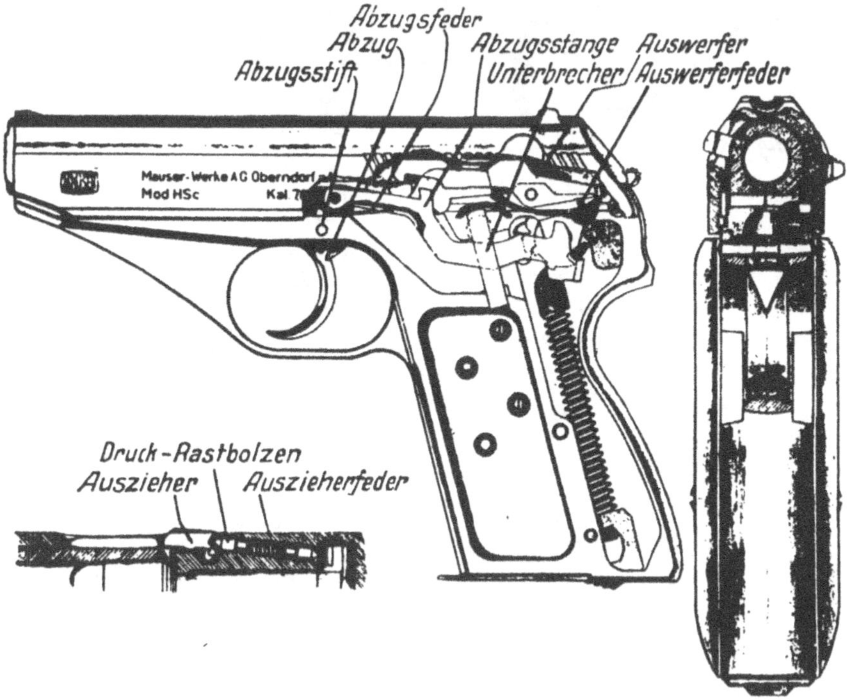

MAUSER MODEL H SC. CALIBER .32

Left side view with side cutaway to show arm uncocked but ready for firing by double action pull-through on trigger. When there is a cartridge in the firing chamber, it is not necessary to thumb or slide cock this weapon to fire it for the first shot.
Rear detail.
Slide detail showing extractor and spring; and thumb safety spring.

barrel against the abutment near the barrel chamber. This compression provides the motive power for forward motion of the breechblock-slide. The slide is machined in the rear breech section to permit it to thrust back and ride over the hammer. The hammer rotating on its pin thrusts its shaft (or strut) down, thereby compressing the mainspring mounted around it. The sear catches and holds the hammer at full cock. The magazine spring forces a cartridge up in line for feeding as the slide is pulled back far enough to clear the top of the magazine.

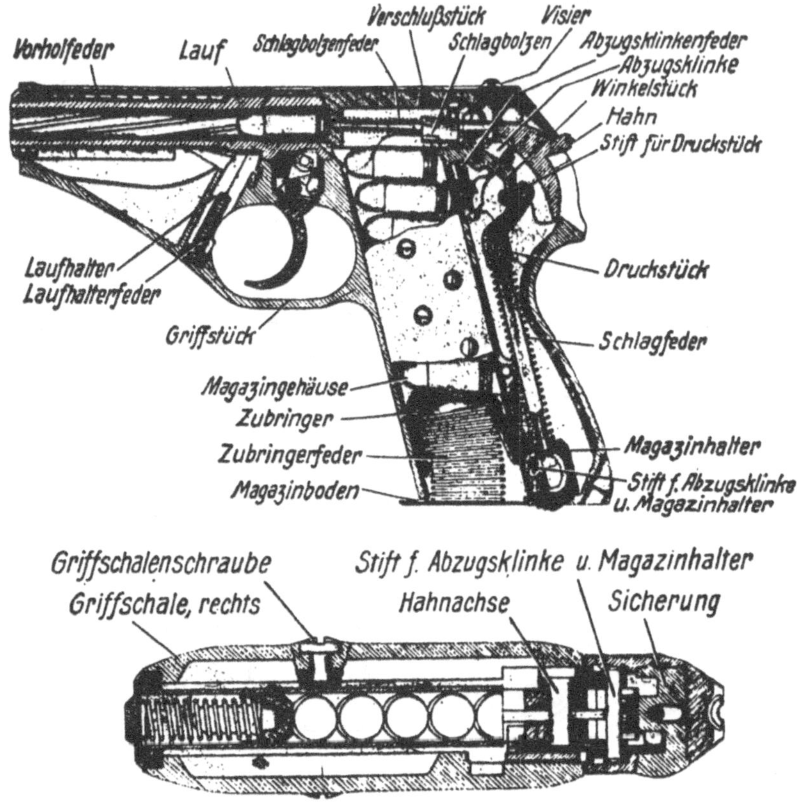

MAUSER MODEL H SC.
Left side view showing chamber loaded and all parts at rest.
Rear elevation showing arrangement of cartridges in magazine.

When the grip on the breechblock-slide is released, the recoil spring expands and thrusts forward on the head of the slide to drive it forward. The feed surface on the bottom of the breechblock face of the slide hits a cartridge in the top of the magazine and forces it into the firing chamber. The extractor through action of its spring transmitted through the buffer pin snaps the head of the extractor into the extracting groove in the cartridge case. The hammer is held back by a sear member called in this design the "trigger catch."

If it is desired, any time the hammer is down in this weapon and the safety is off, the hammer may be pulled back by the thumb for cocking exactly as in a standard revolver. The advantage of this is that when the firing chamber is loaded, while a straight pull-through on the trigger will fire the arm, thumb cocking the hammer when there is time affords a lighter and smoother trigger pull, permitting better aiming for the first shot.

Preparation for firing with the safety applied. When the thumb safety is applied, the pistol is loaded for firing exactly in the same manner as when the safety is off. However, when the breech slide moves forward, the hammer cannot remain cocked. It will travel forward safely with the slide to closed position.

The hammer in moving forward cannot strike the firing pin. When the safety is turned, the cam surface on the safety shaft working against a cut in the firing pin suitably machined, pulls the firing pin further inside the breechblock away from the blow of the hammer, and raises its rear. Thus, when the safety is applied, a cylindrically shaped lock on the firing pin enters a hole in the breechblock which positively prevents the firing pin from moving in its longitudinal direction.

It is impossible to cock the hammer when the safety is on because the shaft of the safety locks the trigger catch (a sear-type member), As the slide is pulled back, it of course thrusts the hammer back and rotates it to the rear compressing the mainspring. However, as the safety is in the breechblock section of the slide, its shaft automatically depresses the trigger catch during forward motion, thereby preventing the hammer from staying at cock. This trigger catch is shaped as a lever.

The value of this system is that it is possible to load the firing chamber by pulling the breechblock-slide back over a loaded magazine with absolute safety. When it is not intended to use the arm immediately, the safety should be applied *before* the chamber is loaded. This device serves the further purpose, moreover, of permitting the hammer to be safely lowered on a loaded chamber. In the typical hammer-type automatic pistol, lowering the hammer on a loaded chamber can be dangerous, for if the hammer slips it may fire the cartridge. If the pistol is the enclosed hammer (called "hammerless") type, it is impossible to lower the hammer on the firing chamber when loaded in most designs. (The Sauer Model 38 Pistol is an exception to this rule).

With this type of construction, when the chamber is loaded, the arm may be safely uncocked by merely pushing down on the thumb safety lever. This will (a) withdraw the firing pin within the breechblock away from the face of the hammer stroke, (b) lock it securely against longitudinal motion, and (c) then depress the trigger catch to let the hammer fall safely against the face of the breechblock slide.

The Action Which Takes Place During Firing

The gas pressure forcing back against the head of the cartridge case as the arm is fired, transmits the blow sharply to the face of the breechblock against which the case head rests. The breechblock-slide starts back in its guides in the receiver. The extractor carries the empty cartridge along with it until it hits the ejector and is thrown out the ejection opening in the right side of the slide.

This rearward motion of the slide compresses the recoil spring, cocks the hammer and compresses the mainspring as already outlined. As the slide moves back, the smooth face on its left rear bottom surface rides over and presses the interruptor down. (When the slide is closed and the arm ready for firing, the tip of the interruptor is nested in a notch in the underside of the slide on the left.) This pulls the trigger sear away from the appendage of the trigger catch. The trigger catch, thereby freed from trigger contact, prevents the firing of more than one shot. Pressure of the trigger catch spring forces the catch into the hammer holding it in cocked position.

The recoil spring carrying the slide forward after the completion of its rearward stroke, takes the top cartridge from the magazine and chambers it.

The extractor claw enters the groove of the cartridge case and the extractor protrudes about 1 mm outside the surface of the slide. This protrusion may be seen or felt. It is a positive indication *that the firing chamber is loaded.*

When pressure of the trigger finger is released momentarily the trigger spring moves the trigger forward and remakes contact between the interruptor, the trigger sear, and the trigger catch. Thus a pull on the trigger is transmitted to release the trigger catch, (or sear) from its grip in the hammer and permit the firing of the next cartridge.

This process is repeated automatically whenever the trigger is pulled. When the last cartridge has been fired, the magazine follower catches the breechblock and holds it fast. Thus the slide remains in rear position to indicate that the pistol is empty.

As the empty magazine is withdrawn, the ejector swivels into the place of the magazine platform to prevent forward motion of the breechblock-slide. Inserting a loaded magazine, then drawing the breechblock back far enough to free it from the ejector, will permit it to be released and to be drawn forward by the recoil spring to chamber a cartridge ready to resume firing instantly.

Closing and Releasing the Safety

Both these operations are performed manually. When the thumb of the right hand pushes the ribbed extension piece of the safety lever upwards and forward, a red spot is uncovered on the side to indicate that the safety is "off."

When the ribbed extension piece covers this red spot, the safety is "on." Note that the hammer cannot be kept cocked with the safety on as already described, and that the pistol *cannot be disassembled* until the safety is "off."

Disassembling

Push the magazine catch and withdraw the magazine. Then pull the slide back to eject a cartridge, if any, in the firing chamber. Put the safety in "off" position. Cock the hammer. Then put the safety "on" again. Take the pistol in the left hand and with the thumb of the left hand press down the barrel catch or holder in the front of the triggerguard and at the same time with the right hand push the slide forward and remove it by lifting it off. Holding the slide in the left hand, with sights down and muzzle outwards, with the thumb and index finger of the right hand press the rear end of the barrel about 3 mm forward and then remove it together with the recoil spring by an upward lifting motion.

When the safety is "off" and the hammer is uncocked, this pistol cannot be disassembled. To attempt it will be to injure it. This is important to remember if you own one of these arms.

Reassembling

When reassembling this pistol, follow the dismounting procedure in reverse order. First, put the safety "on" and cock the hammer.

Then insert the barrel and the recoil spring around it with a slanting forward motion into the slide. Hold the receiver in your left hand and put the slide (with the barrel and recoil spring in place in it) on the receiver guides from above with light pressure so that the front guide studs on the barrel fit into their corresponding recesses in the receiver. (Retract the breech a trifle and then push forward to make sure it locks.

37. UNCOMMON MAUSER PISTOLS

Like all other arms manufacturers, Mauser constantly experimented. Occasionally experimental models of pistols manufactured at that plant and returned to this country by servicemen, will be encountered. In general these will be hand manufactured weapons of which only one or two were made and which did not prove successful enough to warrant manufacture.

The one exception to this is a weapon manufactured in 1913 along the general exterior lines of the Mauser Model 1910 Pocket Pistol but which was constructed with a special breech mechanism and which uses the 9 mm Luger cartridge.

According to official Mauser records, this arm was designed for use by Russian officers. A Brazilian Army Commission also evidenced interest in this design.

This pistol was never put into major production, however, and only a few hundred pieces were manufactured. The factory turned its entire production over to the manufacture of arms for Germany, of course, during World War I. Manufacture of this pistol was never revived.

Occasional samples will be encountered, since a few together with their parts were sent to the Stoeger Arms Company in New York, for sale in this country.

This weapon is *not* equipped with a positive lock of the type of the Mauser Military pistol. However, it has a *delaying* system which differentiates it from the strictly *un*locked blowback type of Mauser pistol.

There is a release catch provided ahead of the triggerguard which controls two flat steel locking pieces, pivoted in the front part of the receiver ahead of the triggerguard where they are retained under powerful spring pressure, constantly tending to thrust them inward.

To load the firing chamber, it is necessary to release this powerful spring arrangement by pushing in on the release catch, before the slide can be drawn back by hand.

The slide has a projection machined at its forward end extending into the recess in the receiver suitably beveled at its rear face, to act upon and cam the dual locking strips outward against the pressure of their springs as the slide starts to the rear.

At no time is this slide *locked* to the barrel. However, since the slide in its rearward movement must overcome not only the pressure resistance of the customary recoil spring and the inertia of the parts to be moved to the rear, but must also overcome the side thrusting pressure of the spring-supported steel finger pieces, the additional force necessary to overcome these added resistances serves to provide a hesitation which permits use of the powerful Luger 9 mm cartridge *when the pistol is in good operating condition.* The recoil is considerable however.

This arm was designed to be fitted with an adjustable shoulder stock holster as in the standard Mauser Military Pistol and has an adjustable rear sight for long range shooting.

The manufacture and machining of this arm and all its components are of the very highest order. The arm itself, however, is essentially a collector's weapon.

38. LUGER AND P-38 PISTOLS MANUFACTURED BY MAUSER

These arms are in every way the counterparts of the earlier Lugers made by Deutsche Waffen und Munitions Fabriken whose facilities were taken over by Mauser after World War I.

The Luger pistol as manufactured by Mauser before World War II may be identified by the standard Mauser name and trademark which appears on the toggle on top of the pistol.

Until 1937, Mauser manufactured these pistols only in the 7.65 mm (.30) Parabellum or Luger type. No legal manufacture of any caliber larger than this was permitted in Germany under the Treaty of Versailles rules, during that period. Mauser pistols of this time made on the Luger design have 3.75 inch barrel, 8.25 inch overall length, and magazine capacity of 8 cartridges. In 1937, Mauser provided these pistols with wooden stock similar in design to those furnished with the Mauser Military Pistols. These stocks were hollow and were used to carry the pistol when not in use. The pistol could be attached by a mortise through the forward end of the stock and used as a carbine. This model had a barrel about 8 inches long.

As the war approached, Mauser again manufactured standard Luger pistols of 9 mm caliber with 4-inch barrels. Luger pistols manufactured during World War II for the German Government by Mauser may be identified by the date and by the code stamp "byf".

The Luger Pistol

The Luger (or Parabellum) pistol is a short-recoil operated, locked breech weapon. The breech is securely locked until the bullet is out of the barrel. Since no gas can escape, a uniform pressure is maintained behind the projectile.

The locking system operates on the toggle-joint lever system much in the fashion of a human knee.

During the locking period, the toggles lie below the line of thrust. When the barrel reaches its full recoil position and is halted together with the part of the sub-receiver into which it is screwed, and with which it travels in the main receiver guide, the toggles hit ramps and are buckled up, pulling the breechblock directly back away from the face of the chamber to extract and eject the empty case. During this buckling movement, a bell-crank lever arrangement attached to the rear of the toggle is pulled up to act to compress a coil spring mounted around the guide in the rear of the grip section of the receiver. At the end of the recoil stroke, this spring reasserts itself and pulling down on the bell crank thrusts the toggle straight ahead. The breechblock strips the cartridge from the top of the magazine and chambers it. The toggle arms are flattened out so their joint is lower than the line of the breechblock, thereby providing positive locking support.

This is a striker-fired weapon, the striker being cocked by an arm on the toggle-joint engaging in a projection on the striker pin to pull it back as the toggle buckles. The trigger, working through an angle lever in the trigger plate transmits pressure on the trigger along the side of the receiver to force the sear out of engagment and permits the striker to be driven forward by its spring to fire the cartridge in the chamber. The magazine is of standard construction except it is provided with a slot in the right side wall in which travels a button protruding from the magazine follower. Pulling down on this button draws the follower down and permits easier loading than is possible with the standard types of magazine where the cartridge itself must be thrust down to force the follower down.

No grip safeties are provided on Mauser-manufactured Lugers. When the thumb safety is pushed back, it thrusts a forward arm up to block movement of the sear, and prevents firing.

The ejector, which is mounted on top of the breechblock, rises above the face of the breechblock when the chamber is loaded.

Thus, if the projection can be felt or seen, the chamber is known to be loaded.

When the last shot has been fired, this arm stays open, the magazine follower button acting to push a catch up to hold the breechblock. When a loaded magazine is inserted, pulling back on the toggle will free the breechblock from the catch. Releasing it will permit the recoil spring operating through the bell-crank lever to close the action and chamber a cartridge ready for firing.

Dismounting this pistol is very simple. First the toggle is pulled back and up over an empty magazine. The breech will stay open. Then the magazine release button on the left side of the receiver behind the trigger is pushed. This permits withdrawal of the magazine from the bottom of the pistol. The locking catch on the left side of the receiver above the triggerguard is then pushed down. Pulling back on the toggle and easing it forward will force the side plate above the trigger out. This plate with the attached trigger lever is removed. The pistol is then turned upside down to prevent the bell-crank lever from causing dismounting trouble, and the barrel and breech assembly are slid off the main receiver. The trigger and its spring may be pulled out of the receiver if necessary. If the toggle is buckled slightly to take pressure off its connecting pin at the rear, the pin may be driven out from right to left, and the breechblock and toggle assembly may be pulled to the rear out of their guides in the sub-receiver section into which the barrel is screwed. All other parts may now be dismounted without difficulty.

The characteristics of the smaller Luger cartridges are as follows: Caliber 7.65 mm. The lead bullet is full metal jacketed as a rule and weigh 93 grains. (Some ammunition was made with soft or hollow point bullets protruding from a partial jacket). The muzzle velocity with 3.75 inch barrel averages about 1225 feet per second, though it may be 50 feet per second higher or lower in ammunition of varying manufacture. The average striking energy is 310 foot pounds, and the average penetration at 15 feet and ⅞th inch soft pine board is 11 boards. This cartridge when used in the long barreled Luger gave ballistics about ten percent greater. When fired with a shoulder stock as a carbine, this arm is reasonably effective at 500 yards and is dangerous at 1500 yards or more.

The 9MM Luger Cartridge

This cartridge is variously loaded with a bullet averaging 125 grains. The bullet is generally full jacketed but was also manufactured with soft point, and with hollow point. This cartridge was loaded as low as 1025 feet per second velocity (with an M. E. of 340 foot pounds) to as high as 1500 feet per second velocity. In one standard loading producing 1110 feet per second velocity the penetration at 15 feet is 10 soft pine boards of ⅞th inch thickness.

The Luger pistol has been manufactured since 1900. Several German manufacturers have made this arm. It also has been manufactured in small quantity by Vickers in England and to a limited extent in Switzerland.

Arms made and manufactured during the periods of World War I and World War II and many manufactured and built up from spare parts between the wars and sold for export are not of the best quality.

Any Luger pistol bearing the Mauser trademark is of the finest quality both of material and workmanship. However, it must be kept in mind that unless Mauser name or code is found on all principal parts, the arm may be assembled one.

These pistols were manufactured by Mauser right up to the closing of World War II, the war product being inferior in finish and workmanship to the commercial manufacture.

The only difference between the two calibers in general is the bore of the barrel.

39. EARLY MAUSER PISTOLS AND REVOLVERS

Mauser made two patterns of revolvers, both of unusual design utilizing a coil mainspring as shown in the drawing housed below the cylinder in the frame.

The first type was a solid frame revolver. A loading port of standard type was provided on the right side of the frame. This was a standard single action type weapon in which the hammer had to be cocked for each shot.

The second type was a hinged frame design with automatic extrac-

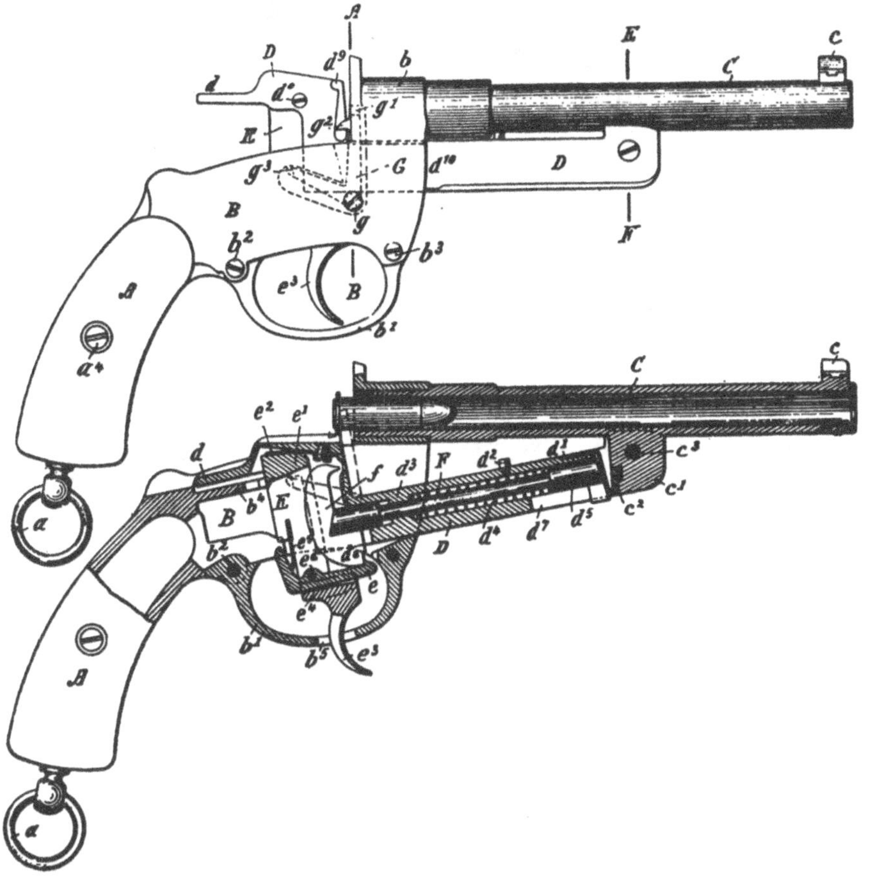

MODEL 77 MAUSER SINGLE SHOT PISTOL.

(1) Right side view, action closed.

(2) Right side view with action open and cartridge inserted in chamber. Note that the act of opening the pistol to load drives back the concealed hammer and cocks the mainspring.

tion. This also was a single action revolver. The act of cocking the hammer caused a ball shape protuberance at the lower end of the hammer to thrust forward a guide and compress the mainspring.

The really unusual feature of this design, however, was the method of revolving the cylinder. This was effected by a zig-zag series of cuts

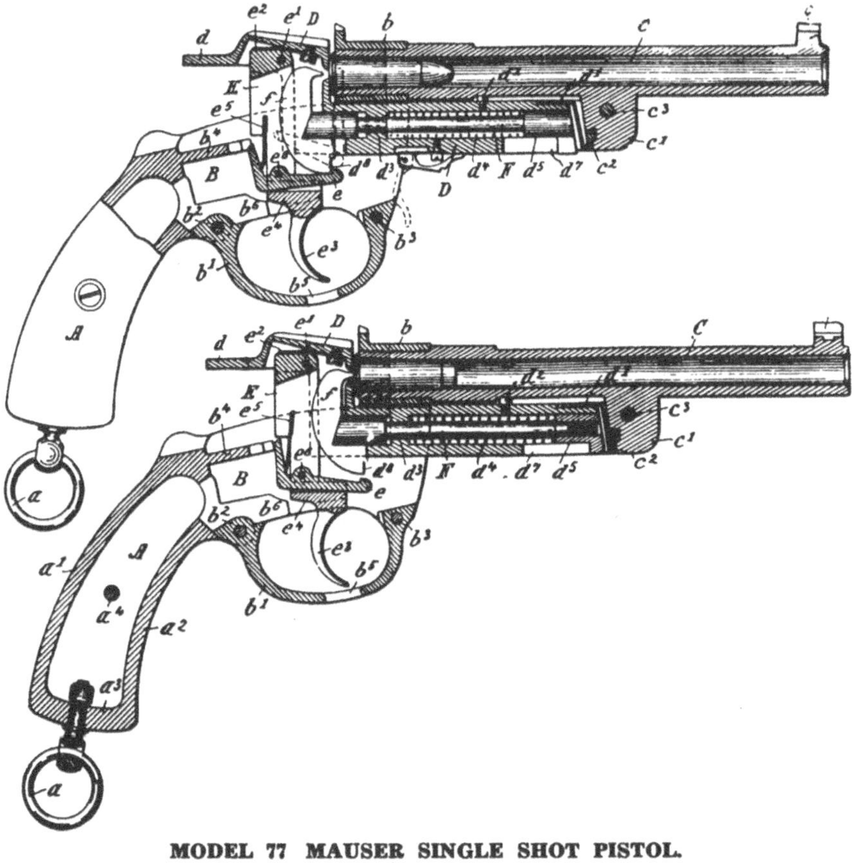

MODEL 77 MAUSER SINGLE SHOT PISTOL.
Pistol loaded and ready to fire. Note the short hammer travel.
Pistol fired, all parts at rest.

in the cylinder wall. During the act of cocking, the stud operating in the cut brought the cylinder around the distance of one chamber and lined the chamber up with the mouth of the barrel.

Samuel Colt developed a cylinder revolving on this general system. It was never produced in quantity, however. The English Webley-Fosbey Automatic Revolver and the U. S. Union Arms Company Automatic Revolver also used modifications of this cylinder system.

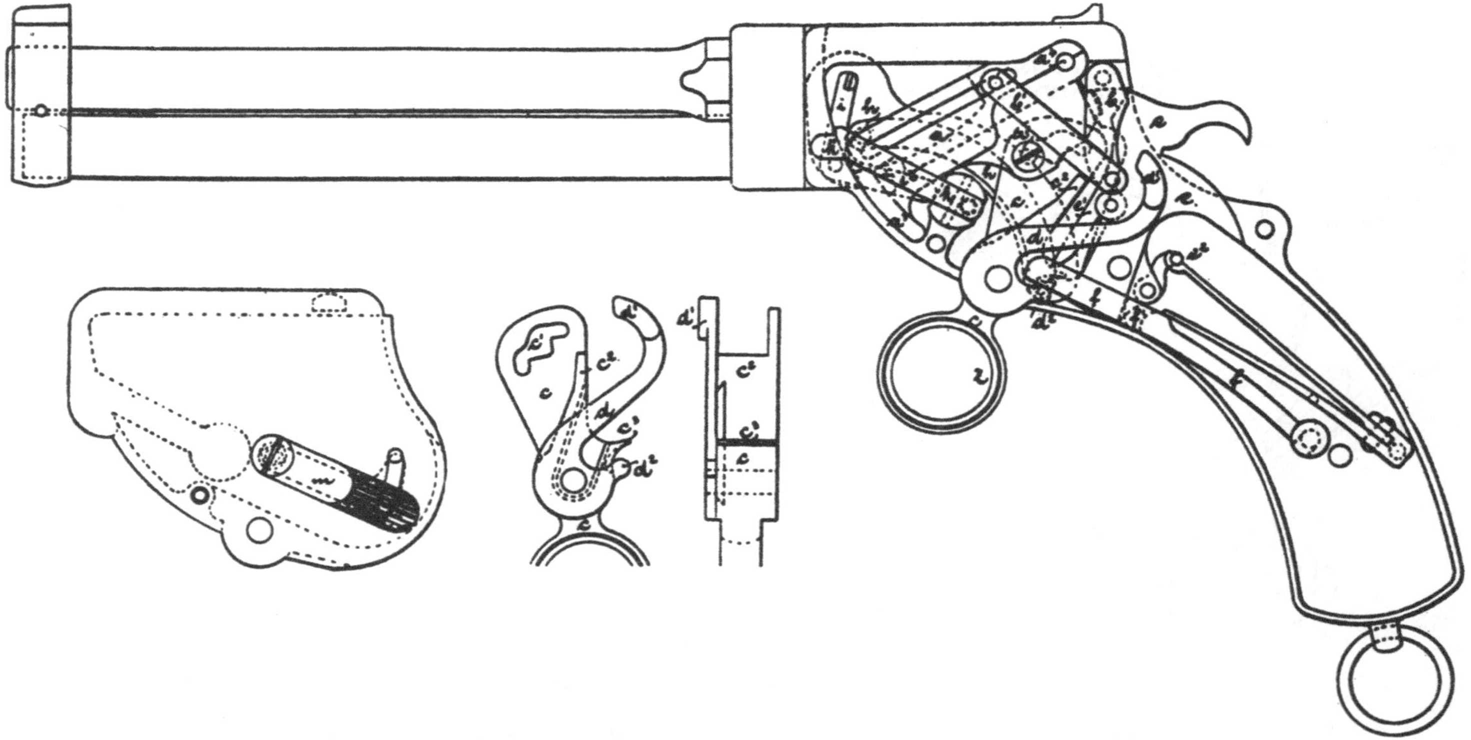

MODEL 86. MAUSER MAGAZINE PISTOL WITH TUBE MAGAZINE AND BLOCK LOCK
Top Line: Details of lockwork, all parts at rest, left side view.
Second Line: Details of safety and lock plate.

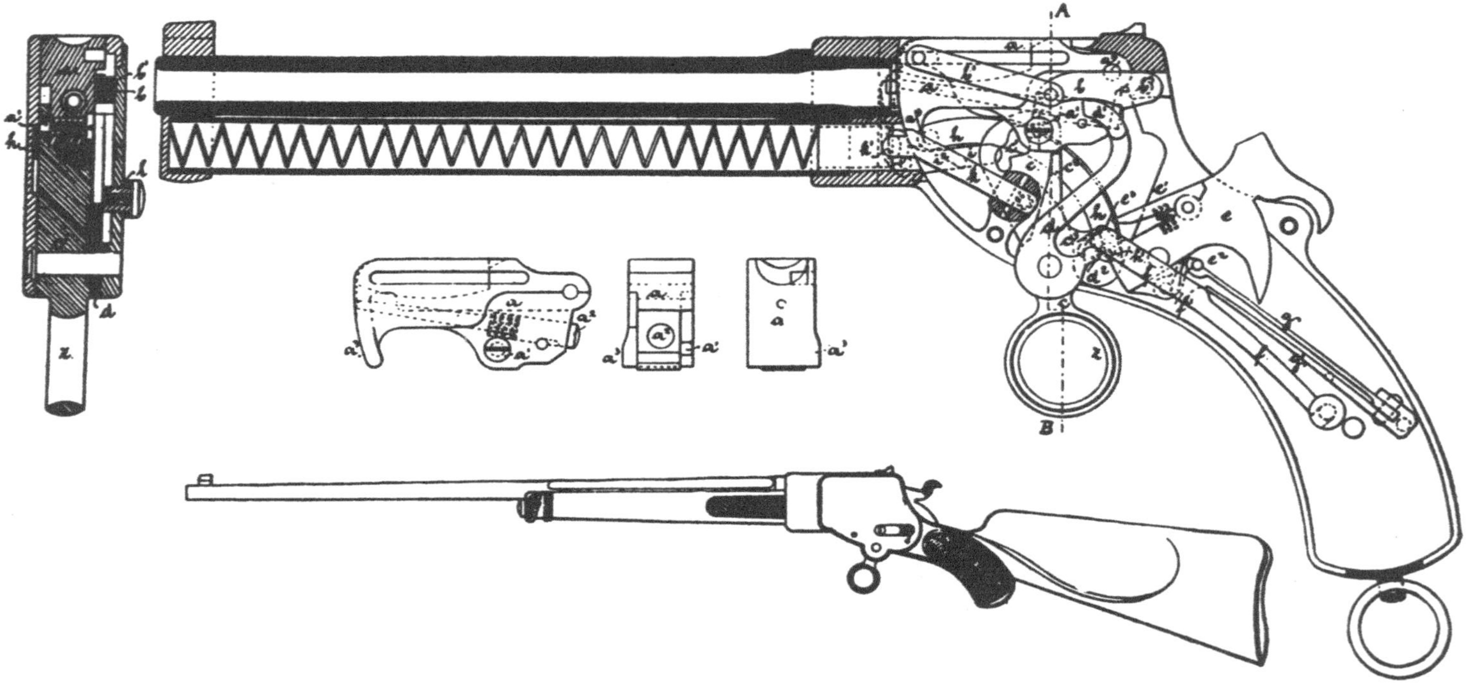

MODEL 86. MAUSER MAGAZINE PISTOL WITH TUBE MAGAZINE AND BLOCK LOCK
Top Line: Left side and rear details. Magazine is empty, but hammer is at full cock and pistol is locked.
Second Line: Left side view of Carbine model of this pistol design.

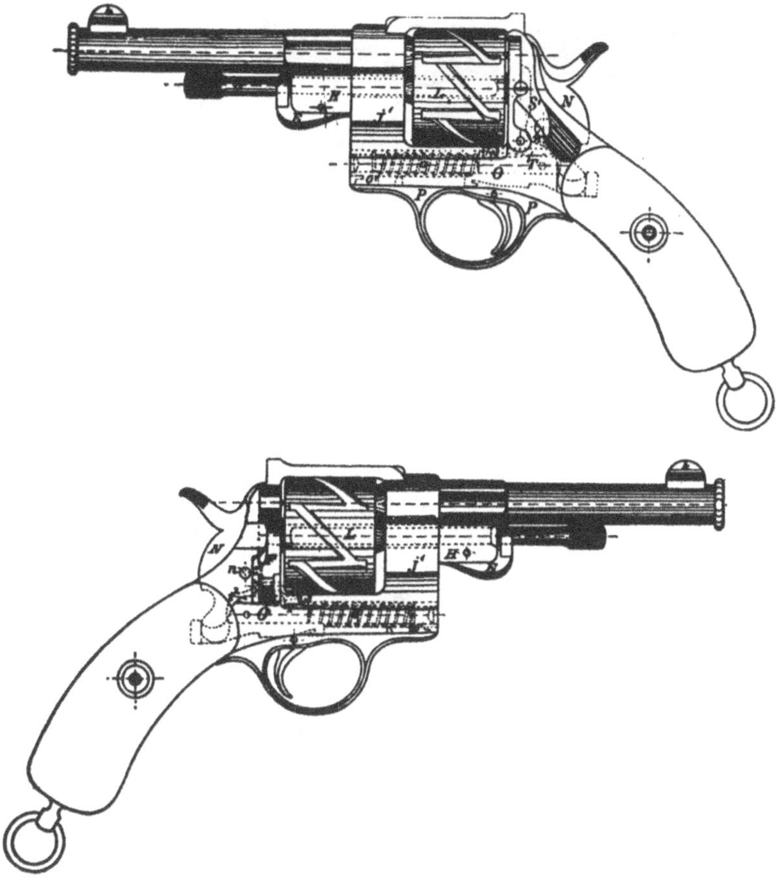

MODEL 78 MAUSER REVOLVERS.

Left and right views of Mauser solid frame revolver. This single action revolver was loaded through a standard type ejection port on the right side of the arm. Note the unusual mainspring and cylinder turning design.

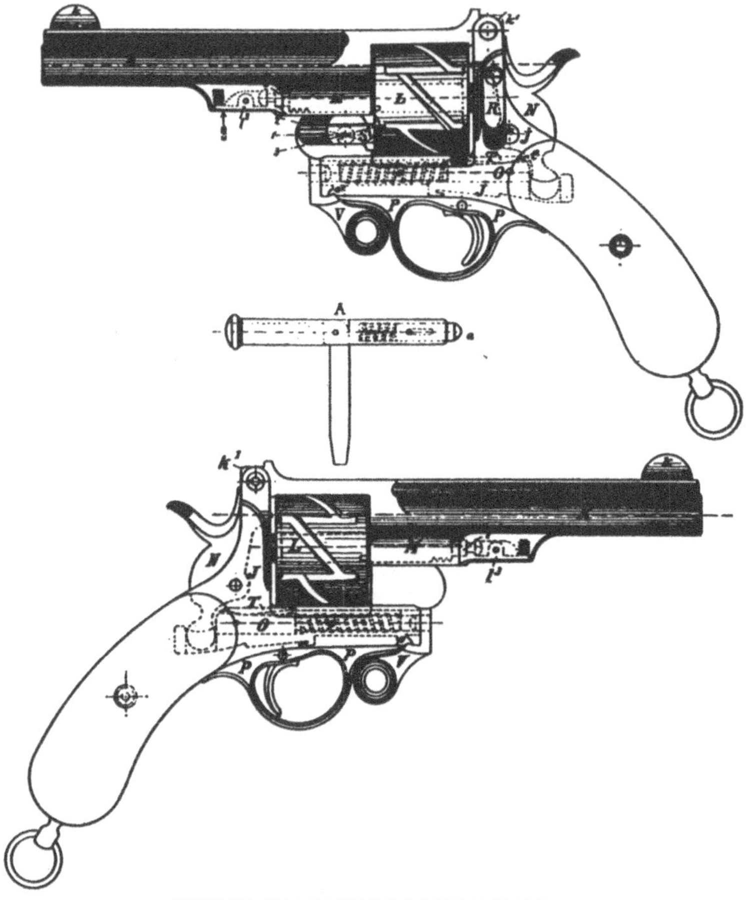

MODEL 78 MAUSER REVOLVERS.

Left and right views of Mauser hinged frame revolver. Turning forward lock (V) permitted the barrel and its strap to be hinged up for loading and for extraction. The firing and cylinder revolving systems are those of the solid frame type. This too was a single action arm in which the hammer had to be thumb cocked for each shot.

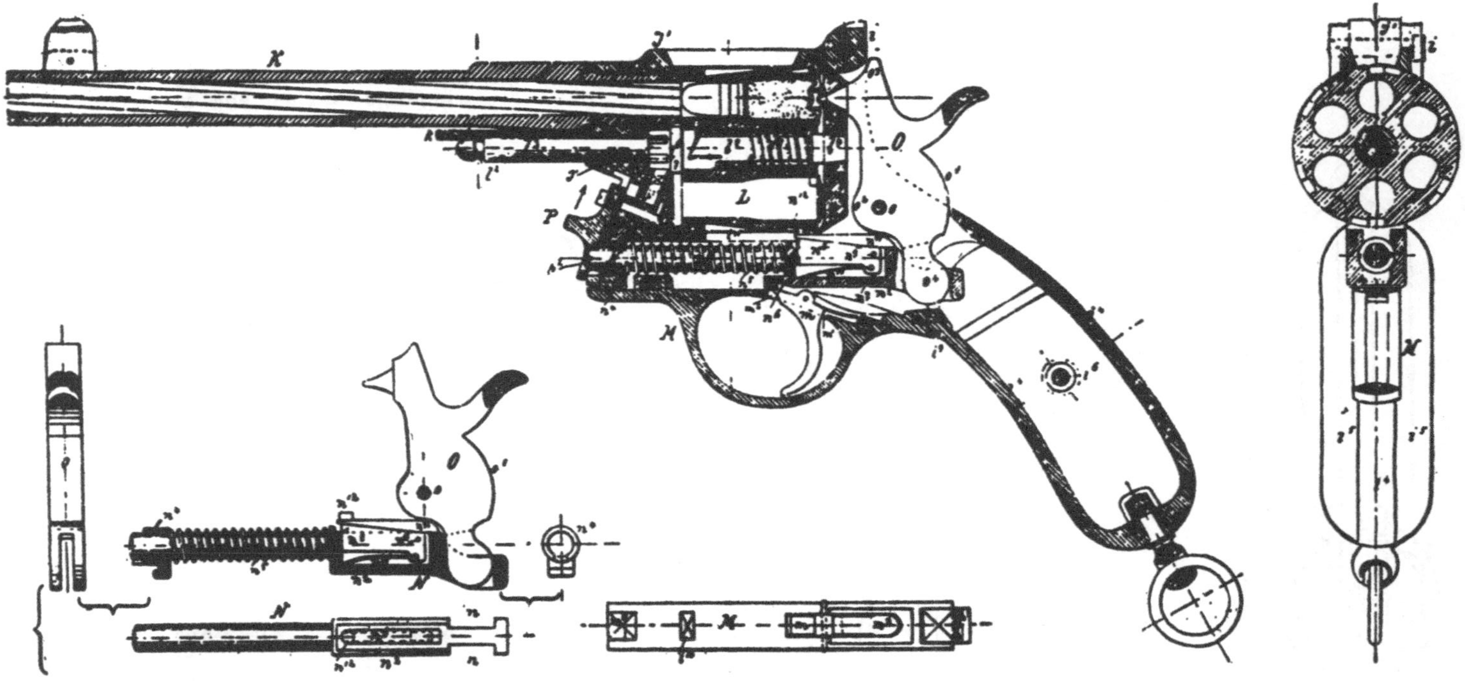

MODEL 78. MAUSER IMPROVED REVOLVER.

Top Line: Left side and rear cutaway views showing details of construction.
Second Line: Details of hammer, mainspring and cylinder revolving mechanism.

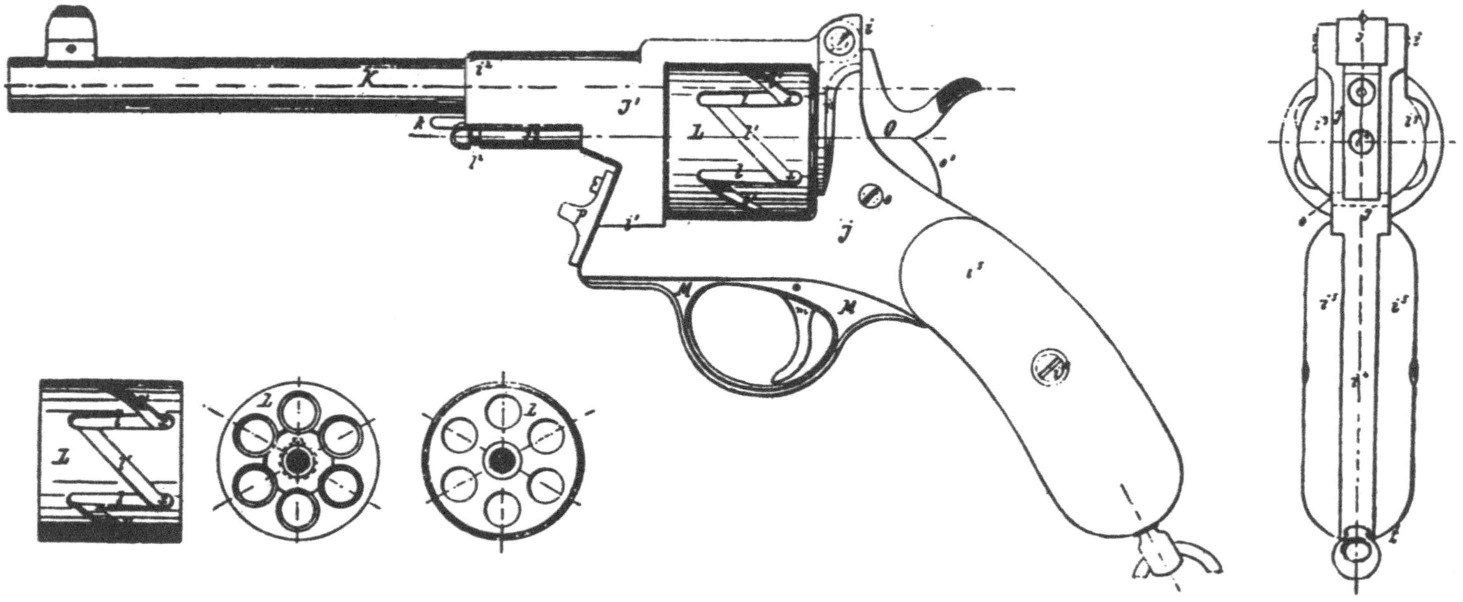

MODEL 78 MAUSER IMPROVED REVOLVER.

Top Line: Side and rear elevations.
Second Line: Side, rear and front cylinder views.

CONCLUSION

Every Mauser rifle, pistol and revolver ever manufactured on a production basis has been pictured or adequately described in this book.

If you encounter a genuine Mauser weapon not listed herein, you have a hand built experimental weapon discarded by Mauser as unsatisfactory; and probably brought here by one of our soldiers who picked it up after the Mauser factory had been visited by our troops.

If you have a Browning, Colt, Savage or similar pistol with the Mauser trademark on it together with the actual maker's name, you merely have a weapon originally bought by Mauser for study and marked with its property stamp. Mauser didn't make it.

This book is the *only* source of data in English on the history of Mauser, and it goes far beyond any data ever published in German.

Where dimensional data is not provided, as in the case of receiver lengths, it is because Mauser specifications were based on special steels and known cartridge ballistics; to publish this data generally might result in its improper use by unqualified gunsmiths.

Americans readily understand that the Springfield today differs so little mechanically from those made 40 years ago that the description of one essentially covers all.

Curiously, many Americans fail to realize that the same thing applies to Mauser 98 weapons. Every attempt has been made herein to show the very minor internal differences in Mauser rifles of various periods and calibers. The receiver bridge may be squared off as in some Mausers made for export; sights differ; bolt handles have different shapes; sling positions vary; lengths and weights differ; magazine releases are of button, stud or lever type. But the rifles themselves are nearly alike. Except that after 1924 the gas flanges on the bolt plugs are larger and the magazine followers hold the bolt open when the magazines are empty, any design differences in the series from 1898 to 1944 are those connected with simplification of manufacture. Since such minor modifications vary with every plant which ever made Mauser type actions or parts, it is not practical to list them.

The ammunition for Mauser and Mauser System rifles could not be more than touched upon in a book of this size and type. Those interested in even sampling the range of foreign cartridge data are herewith referred to the DWM and RWS catalogs which, while they contain an enormous amount of data, do not even scratch the surface. Thousands of foreign gunsmith built rifles on Mauser and Mauser type actions, and a summary of the caliber combinations or rifle specifications is as difficult as it is useless.

Finally, the reader's attention is called to the fact that many of the illustrations used were made by Mauser years ago; a fact which may interfere somewhat with reproduction, but which provides an historical guarantee of authenticity.

Smith & Wesson Hand Guns....................................McHenry & Roper
Hunting Pronghorn Antelope....................................Bert Popowski
The Woodchuck Hunter..Paul Estey
Small Bore Rifle Shooting..Edward Crossman
Modern Bow Hunting...Hiram Grogan
Salt Water Fishing Is Easy..Jerry Sylvester
Fishing Flies and Fly Tying.....................................William Blades
This Business of Exploring.............................Roy Chapman Andrews
Handbook for Shooters and Reloaders.......................P. O. Ackley
Greatest Fishing...Joe Brooks
Hunting Our Biggest Game.......................................Clyde Ormond
How To Hunt Deer...Edward Freeman
Burning Powder...Douglas B. Wesson
Practical Rifle Shooting...Walter Winans
Handloading Ammunition..J. R. Mattern
Shooting...J. Henry Fitzgerald
Rifles & Rifle Shooting...Charles Askins
Mauser Rifles and Pistols..W. H. B. Smith
The Book of the Pistol & Revolver............................Hugh B. C. Pollard

Made in United States
Cleveland, OH
19 September 2025

20603893R00136